NANOFUTURE

NANOFUTURE
What's Next for Nanotechnology

J. STORRS HALL, PhD

Chief Scientist of Nanorex, Inc.,
Fellow of the Molecular Engineering Research Institute

Foreword by K. ERIC DREXLER

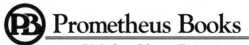
Prometheus Books

59 John Glenn Drive
Amherst, New York 14228-2197

Published 2005 by Prometheus Books

Inquiries should be addressed to
Prometheus Books
59 John Glenn Drive
Amherst, New York 14228–2197
VOICE: 716–691–0133, ext. 207
FAX: 716–564–2711
WWW.PROMETHEUSBOOKS.COM

09 08 07 06 05   5 4 3 2 1

Library of Congress Cataloging-in-Publication Data

Hall, J. Storrs.
    Nanofuture : what's next for nanotechnology / J. Storrs Hall
        p. cm.
    Includes bibliographical references and index.
    ISBN 1–59102–287–8 (alk. paper)
    1. Nanotechnology I. Title.

T174.7.H35 2005
620'.5—dc22

                                                    2005001789

Printed in the United States of America on acid-free paper

for Purser Hewitt, 1905–1986

CONTENTS

FOREWORD

The projected impact of nanotechnology has been touted as a second industrial revolution. Not third, fourth, or fifth, because despite similar predictions for technologies such as computers and robotics, nothing has as yet eclipsed the first.

The original industrial revolution transformed our way of life. At the level of the individual it doubled average lifespan; at the level of national affairs it made possible truly global civilizations. Will nanotechnology measure up on that scale?

Nanotechnology already encompasses remarkable inventions, and yet more are in the labs and on the computer screens of theorists. How are we to judge whether any set of new inventions, however ingenious, will change the world beyond some small technological niche? Read this book and you will see.

Reaching a solid understanding of new technology—the understanding necessary to judge its effects—is an intellectual adventure. I could not wish you any better guide on such a journey than Josh Hall.

Before the term *nanotechnology* had reached a tenth of its current popularity, he had already formed the first worldwide Internet discussion group and led the discussion for a decade. He has done research and development in nanotechnology since the early days, with multiple inventions and discoveries to his credit.

More important than his scientific credentials, however, is the broad context he brings to his explanations. Reading this book, you will learn not only about nanotechnology, but about all technology—

and not as a mere collection of gadgets, but as part of the structure of the world. It wasn't just the power of the steam engine that made the industrial revolution; it was the relationship between *what it could do* and *what needed to be done.*

At the height of the Roman Empire, Heron of Alexandria used steam to power an engine, yet this remained a mere toy. It launched no revolution. It lay forgotten until the context was right, and then steam power changed the world.

It may be difficult to accept that profound changes in technology can reshape our world, but a sense of historical context can help. Try to imagine the history and concerns of the twentieth century without its great technological advances: antibiotics, automobiles, aircraft, radio, television, motion pictures, electric lights, washing machines and dryers, indoor plumbing, and computers. War and peace, poverty and prosperity, dreams and nightmares—all would be different.

The pace of technological change is accelerating, and nanotechnology will be central to that change over the coming decades. The next few decades may well bring more change in technological capability than the past century. The lessons of past technological revolutions are our best guide as we face the next.

History, properly told, is first and foremost a story. Technological history is a story of human needs and physical possibility. Technological forecasting must tell a similar story of human needs fulfilled within the possibilities set by physical law. To understand the impact of nanotechnology, we must consider what it makes possible, what needs it fulfills, and how necessity and possibility together have shaped our world before.

You'll get the whole story here.

K. Eric Drexler
Los Altos, California
August 2004

CHAPTER 1

WHAT IS NANOTECHNOLOGY?
And Why Is There
So Much Confusion about It?

Who among us would not be glad to lift the veil behind which the future lies hidden; to cast a glance at the next advances of our science and at the secrets of its development in future centuries?

—David Hilbert,
Second International Congress of Mathematicians,
Paris, 1900

The first half of the twentieth century witnessed an explosion of technology that deeply affected the way we live. In 1900, heavier-than-air flying machines were widely believed impossible; in 1950, jets were approaching the speed of sound. In 1900, most people did not have cars, electricity, or indoor plumbing. By 1950, they did. The same fifty years saw substantial parts of the development of antibiotics, radio, television, plastics, nuclear weapons, and the computer. Tractors, harvesters, and similar equipment cut the number of people required to produce a given quantity of food by a factor of ten.

To a great extent, the march of progress continued through the second half of the century. Jet airliners became common, and with the Boeing 747, air freight became economical for some kinds of goods. Televisions, computers, cell phones, and similar gadgetry became ubiquitous. The global communications network along with the construction of enormous freight ships and tankers wrought a world economy more integrated at the turn of the twenty-first century than the national one had been at the turn of the twentieth. Men walked on the Moon.

And yet somehow the grand promise of technology seemed to lose its magic. The footprints on the Moon have lain undisturbed for decades. In the late twentieth century, Western civilization produced an artifact with a volume surpassing the Great Wall of China. It was the Fresh Kills landfill, New York City's garbage dump on southern Staten Island. Automobiles don't go very fast when the roads are jammed full of them. Even the production of food in unheard-of quantities resulted in an epidemic of obesity and heart disease.

The shift in perceptions is complex but has a few main roots. First is simply human nature: the promise of technology in the early twentieth century was largely fulfilled, in the industrialized nations at least. People who are well fed, warm, and not faced with hard physical labor turn their attention elsewhere. We did not have two global wars in the second half-century as we did in the first. War has a tendency to focus attention toward the means of victory, and away from undesirable side effects.

If you had told a farmer in 1900 that his descendants a century thence would spend their time mostly inside heated and air-conditioned (air what?) buildings, sitting on cushioned seats, talking, reading, and writing, he might have believed you if he was a believer in Progress. If you had told him they would call this "working," he might have laughed in your face. The triumphs of twentieth-century technology—full stomachs and the lack of having to do backbreaking labor—are taken for granted, and lesser problems have been elevated in their stead.

And finally, there is a real phenomenon of diminishing returns. A safety razor with two blades is not twice as useful as one with only one blade. Cars hit a speed limit, and thus a usefulness limit, because of the limitations of human reflexes, not limitations of mechanical capability. Airliners hit a speed limit involving economics and optimal flight regimes. The technology for supersonic passenger flight is there, it just costs too much to pay its way.

In science, the first half of the twentieth century saw a vast flowering of knowledge in chemistry, as physics revolutionized the way we understand ordinary matter with quantum mechanics. It's reasonable to say that a scientist in 1950 could give a good, fundamentally sound explanation of ordinary situations, such as why materials have the properties they do. The scientist of 1900 would have had to do a lot of fudging. He could have told you that snowflakes have sixfold symmetry because the water molecule has a certain shape. The scientist of 1950 could tell you why the molecule has that shape.

What about the latter half of the century? Did science hit the same apparent law of diminishing returns as did technology? In some areas, yes. Physics, after its brilliant success, moved on to mostly more esoteric phenomena, having less to do with the everyday world. The fact that protons, neutrons, and so forth are made from quarks is a marvelous systematization of subnuclear physics, but it has no practical impact on anyone's life. But something else did. Without nearly as much fanfare as it deserved, science as a whole proceeded to crack open the greatest, most essential mystery of all time: the nature and mechanism of life itself.

The story really begins in the early 1600s. In 1633, around the time Galileo was being condemned by the Church, René Descartes wrote a book called *De Homine*[1] in which he tried to explain some of the phenomena of the human body in mechanistic terms. Descartes is famous for the doctrine of dualism, the claim that there is a distinction between mind, which acts mysteriously, and matter, which works mechanically. Vitalism, belief in a "life force" that doesn't obey the mechanistic laws of physics and chemistry, had a strong history in intellectual tradition and an even stronger presence in people's everyday understanding of how things worked. Descartes stuck a foot in the door. He opened the way to seek mechanistic explanations for the phenomena of life. Dualism, intentionally or not, formed the perfect shield against the reactionary and religious forces of the day, allowing one to seek mechanisms for hand-eye coordination, for example, while leaving spirituality, morality, and consciousness to the category of "mind."

Fast-forward to 1825. Chemists were still discovering new elements—the basic rules and properties of how they combine are still being worked out—and they divided the world into two distinct kinds of "stuff": inorganic and organic. Never the twain shall meet. Organic compounds were made by living things, while inorganic ones could be made by chemists in test tubes. You could break organic compounds down, of course; just burn them, for example. But putting them together required some ineffable life force that was universally and seriously believed to exist. Then a young chemistry teacher named Friedrich Wöhler produced an inert solid from a mixture of aqueous hydrogen cyanide and ammonia in 1825, subsequently making the same chemical in different ways from cyanide and ammonia salts. By 1828 he had managed to show that his compound was chemically identical to urea, an organic compound found in urine.

By the twentieth century, progress in everything from physiology to chemistry had made it reasonable to think that all of life was basically mechanistic. However, it was still largely a matter of faith either way. The mechanisms weren't known. By 1944, though, things had advanced to the point that Erwin Schrödinger could write *What Is Life?*[2] This landmark essay made a persuasive case that all the phenomena of life at the cellular level could be part of an unbroken chain of explanation going all the way down to quantum mechanics (of which Schrödinger was one of the main developers). Nine years later, Francis Crick and James Watson discovered the structure of DNA, and the floodgates opened.

When I was in grammar school, my biology teacher explained that four properties made living things different from nonliving ones. These were Organization, Metabolism, Reproduction, and Irritability. Organization meant that organisms have complex inner structure, being made of cells, and cells having a further inner structure. Metabolism meant that they consumed nourishment for growth and activity. Reproduction meant—well, this was a grammar school. And Irritability meant reactivity to stimuli ranging from the conscious actions of humans to the tropisms of plants.

What she didn't explain, partly because it was an introductory course and partly because it wasn't fully understood at the time, was how all those properties are the result of the activities of molecular machines. The key is reproduction. Cells reproduce to form the structure of the body, and ultimately to allow the entire organism to reproduce. A cell is a factory filled with machines and a blueprint. Among the machines are machines to make the machines, called ribosomes, and machines to copy the blueprint, called replisomes. The blueprint itself comes in volumes called chromosomes. Each one is a molecule of DNA.

In the latter half of the twentieth century, we went from having a sketchy notion of how it might work to having our teeth solidly into the details. The "genetic code," deciphered in the 1960s, tells what sequence of amino acids a given piece of DNA is the blueprint for. We can't say, in general, what protein shape that sequence will fold up into, nor the function of the molecular machine the protein forms, but we're getting there. We know a lot of specific cases. A gene is a part of the DNA that's the description of a given machine. We can snip genes from the DNA of one kind of organism and insert them into another, with various results, including glow-in-the-dark pet fish.

All the forms of life we know contain many molecular machines.

The ability to tinker with them, put them together in new combinations, and modify life in various ways, is called biotechnology. As yet, biotechnology hasn't done much in the way of brand-new, invented designs of molecular machines, but that's coming. Biotechnology is limited, however. To begin with, all the machines are proteins. (Biology can build with other materials, creating bone and teeth, but not active molecular machines.) It's stuck with the DNA–RNA–ribosome–amino acid–protein mechanism, and that means it's limited to operating in solution in a narrow temperature range. There are plenty of things it can't make: tin cans, for example.

On the other hand, biology—and biotechnology when we master it—can and does make a staggering variety of mind-bogglingly complex, incredibly adaptable things—creatures—with capabilities our conventional manufacturing technology can't match. We can build Moon rockets, which biology can't, but not insects, which biology does with mad abandon. We can build electron microscopes, which biology cannot, but Joyce Kilmer's dictum that only God can make a tree is still true. A nuclear-powered aircraft carrier, loaded with planes, has something like 100 trillion working parts (most of which are the trench capacitors that store bits in the memories of its computers). So does a housefly.

What if it were possible to combine the capabilities of the biological mechanisms with those of conventional ones? We could have machines that could grow and repair themselves, proliferate without factories, grow from seeds. We could have a technology with the subtlety and adaptability of life, the power of jumbo jets, the efficiency of electric motors, the precision of computers. Is such a technology possible?

It's not only possible, it's on its way. It will change the way we live, what we know, how we think, and maybe even who we are. It is called nanotechnology.

WHAT IS *NANOTECHNOLOGY*?

Nanotechnology: the art of manipulating materials on an atomic or molecular scale especially to build microscopic devices (as robots).
—AOL online dictionary

Nanotechnology: [No entry]
　　　　—*Webster's Third New International Dictionary* (1966)

Most books about future technology tend to follow Clarke's Law: "Any sufficiently advanced technology is indistinguishable from magic."[3] It is difficult to do otherwise: for the general reader, and indeed to the specialist outside his field, much of existing technology is magic. This is not, however, a very useful basis for the reader who wishes to make personal long-term plans, participate in the political and economic decision-making processes, or simply have a feeling of understanding how things work in the world.

If everyone commenting in public about nanotechnology were saying basically the same thing, it would be at least reasonable just to take their word for it. However, there seems to be no widespread agreement on what nanotechnology is, much less what it can do.

One of the first places that the term got wide play was in science fiction. This at least let some of the ideas be discussed, but science fiction is not necessarily a good place to learn about new and important technological developments. In the first place, the science fiction writer is under no obligation to stick to the known, the possible, or even the truth. The demands of telling a good, dramatic story come first. So the science fiction writer will often write the plot first and then cut, stretch, or simply make up the technology to fit.

Here are a few of the things nanotechnology is not: it is not infestations of "nanites" that take over your starship, as in *Star Trek*; it is not evil shape-shifting killer robots from the future, as in *Terminator II*; it is not evil clouds of flesh-eating mites, as in *Prey*. These things are science fiction, with the emphasis on fiction. They are designed by storytellers to have a maximum emotional impact, and have virtually nothing to do with sensible, levelheaded technological prediction.

On the other hand, you'll find people advertising the most mundane advances as nanotechnology: stain-resistant pants, skin creams, oil and paint additives, and so forth, which don't seem significantly different from other cloth treatments, additives, and what have you. Indeed, as I write this, New York's attorney general has been asked to investigate the overuse of the word *nanotechnology* in company descriptions as fraudulent overpromotion.[4] One commentator quipped, "It seems that they're willing to call themselves 'nanotechnology' if their product is made of atoms."

It may help to have a look at the history of the word *nanotechnology* itself. Back in the 1970s, biotechnology was beginning to emerge as a serious possibility with the introduction of recombinant DNA techniques. K. Eric Drexler, an MIT student, became interested in the notion that much of what went on inside a cell wasn't so different, in principle, from what engineers did at macroscopic scales. What's more, it seemed reasonable that, having gained control of the cell's molecular machinery, one could use it the same way that engineers did normal-size machines: making materials, structures, tools, and more machines.

Drexler wrote these ideas in a scientific paper and published it in the *Proceedings of the National Academy of Sciences* in 1981, under the title, "Molecular Engineering: An Approach to the Development of General Capabilities for Molecular Manipulation." Its abstract was as follows:

> Development of the ability to design protein molecules will open a path to the fabrication of devices to complex atomic specifications, thus sidestepping obstacles facing conventional microtechnology. This path will involve construction of molecular machinery able to position reactive groups to atomic precision. It could lead to great advances in computational devices and in the ability to manipulate biological materials. The existence of this path has implications for the present.[5]

During the 1980s, Drexler and a group of friends worked out the possibilities and some of the implications of such a technology. Then in 1987, Drexler published a popular book about it entitled *Engines of Creation: The Coming Era of Nanotechnology*.[6] The people who had been thinking about it and working on it throughout the 1980s had used the word *nanotechnology* all along; but it was the publication of *Engines* that first introduced the term to the public at large.[7]

Engines of Creation is a technophile's dreamscape. It predicts microscopic replicating units able to build skyscraper-sized objects to atomic precision. These could be buildings or they could be spaceships. It discusses artificial intelligence and engineering systems able to handle the enormous complexity such designs would require. It speaks of "easy and convenient" space travel, and describes a spacesuit so light and thin that you almost forget you're wearing it (present-day spacesuits are very awkward and arduous to wear and work in). It mentions cell repair machines and curing "a disease called aging." It talks about cryonics and resurrecting the frozen.

Engines created a sensation in technically oriented circles. There had always been a segment of the population who liked technology for its own sake, or believed it could, and dreamed it would, continually improve the human condition. (This segment is actually smaller now than it was a century ago, when Edison was a popular hero.) These dreams now had a form, with pathways and techniques in sight where only the vague mantra of "technological progress" had been before. They also had a name.

Drexler coined the word *nanotechnology* by analogy to the then already common *microtechnology*, which mostly applied to photolithographic chip-making techniques but was broadly applied to any technology that manipulated matter at the micron scale. Since the new technology would manipulate matter at the nanometer scale, the extension seemed straightforward.

By the mid-1990s, *nanotechnology* had picked up the cachet of a popular buzzword, and started showing up in science fiction, popular scientific, and technical publications—and grant proposals. After all, if *nanotechnology* means dealing with matter on the scale of nanometers, then a huge amount of existing science and engineering was, by definition, nanotechnology: chemistry, molecular biology, surface physics, thin films, ultrafine powders, and so forth. By the turn of the century, computer chips, the original microtechnology, had features small enough that they could be reasonably measured in nanometers. (A micron is 1,000 nanometers, so instead of saying 0.08 microns you say 80 nanometers.)

POPULARITY AND "SUCCESS"

> If you can bear to hear the truth you've spoken
> Twisted by knaves to make a trap for fools . . .
> —Rudyard Kipling, "If"

Now, in a perfect world, a definition would have been an issue for lexicographers. But in reality, the world of science, no less than the world of politics, is driven by fads and buzzwords. And where the two intersect—the science and technology funding agencies—you need a buz-zsaw to get through the buzzwords. So the word *nanotechnology* was rapidly adopted by researchers attempting to get just that one extra edge over their peers in the highly competitive, not to say cutthroat, business of getting government money.

Another thing that funded researchers compete for is the brightest graduate students. Graduate students actually do most of the work in university labs, and for wages that are quite low compared with the standards of the commercial world. Researchers offer a number of nonmonetary rewards, and one of the main ones is for the students to be working in, and learning, new fields at the cutting edge of knowledge. Describing your work as nanotechnology was a good way to make it more attractive to the best students.

So the meaning got stretched, and stretched, and stretched.

And then a funny thing happened. By the turn of the century, there was a National Nanotechnology Initiative, and roughly a billion dollars per year of funding for research under the name. And as will happen, people who wanted the money for other things, and people who just didn't like science and technology in the first place, and jour-nalists looking for an eye-catching story, all went back and read Drexler again—and the first thing they found was the chapter entitled "Engines of Destruction."[8]

Of course, any powerful new technology will bring new dangers; and nanotechnology, being essentially engineering at the level of con-trol that gives life its miraculous powers, is very powerful indeed. So the technophobes, Luddites, and horror fiction writers had a new Frankenstein's monster to play with.

This left the "nanotechnology—stretched version" researchers in an uncomfortable, if ironic, situation. They had adopted a name for their endeavors that was a popular buzzword because of the associa-tion it had with powerful future technologies and revolutionary capa-bilities. But what they were working on was really just more chem-istry, or more ultrafine powders, or what have you. And now they were being attacked for doing "nanotechnology"—this demon's spawn that could destroy everything if the slightest mistake were made. Their immediate, and quite predictable reaction: "Oh, no, nan-otechnology can't do that kind of thing."

And they're quite right; the stuff that's going on in most labs today under the name of nanotechnology may make smaller computer chips, or stronger aerospace materials, or whatever, but it's really more of the same old conventional technology by another name. You don't need to read a whole new book to learn that people are trying to make more stain-resistant (and expensive) pants, or stronger (and more expensive) tennis racquets, or smaller, faster computers. Nor do you need to worry over the fact that marketing departments will be calling these things, and lots of other things over the coming years, "nanotechnology." It's just a word.

There is the appearance of a debate going on in the scientific community as to whether original, Drexlerian nanotechnology could work as advertised. The appearance is deceiving. There are indeed two camps of thought: one is the politically motivated people, largely grant funded, who are worried about the public perception of nanotech dangers impacting their money. They have very much a "not invented here" attitude toward anything even vaguely smelling of self-replicating machines or molecular manufacturing, and have made little effort to understand it. Typically, though not always, you'll find chemists and materials scientists on this side of the question.

The other side is not so politically oriented, doesn't have the money to protect in places where it's vulnerable to every passing wind of public opinion, and—surprise—is not nearly as skeptical. A substantial proportion of scientists and engineers, those without a monetary stake in the technical question, agree with the position I take in this book. Typically, but again not always, you'll find physicists, mechanical engineers, and computer scientists on this side.

Drexler did his doctoral dissertation at MIT on nanotechnology, and then after some further research published it as a technical book entitled *Nanosystems: Molecular Machinery, Manufacturing, and Computation*.[9] No one has ever found a significant error in the technical argument. Drexler's detractors in the political argument don't even talk about it; they tend to avoid technical argument altogether, resorting instead to emotional non sequiturs like, "You're frightening our children."

Some of the potentialities of nanotechnology are indeed frightening. But that makes it even more important to think about it seriously—not to avoid it! It's my considered opinion, after more than a decade of technical study, that nanotechnology as Drexler described it is almost certainly feasible, and nearly as certain to be implemented by somebody, somewhere, within the current century. Given its profound implications for the human condition, it would be incredibly foolish *not* to weigh it carefully in our planning.

WHAT IS NANOTECHNOLOGY?

> But even in the field of science, it is perilous to run counter to the accepted tables of precedence. On no account is it permissible to mention living beings and machines in the same breath. Living beings are living beings in all their parts; while machines are made of metals and other unorganized substances, with no fine structure relevant to their purposive or quasi-purposive function.
> —Norbert Wiener, *God and Golem, Inc.*

So *nanotechnology* really does have two different meanings. One is the broad, stretched version meaning any technology dealing with something less than 100 nanometers in size. The other is the original meaning: designing and building machines in which every atom and chemical bond is specified precisely. I'll refer to the former as *nanoscale technology* when I need to; but I won't refer to it much. The capabilities and dangers of nanoscale technology are simple and straightforward extensions of current trends in the capabilities and dangers of chemistry, materials science, and microfabrication. The majority of new techniques being discovered and trumpeted as the "latest thing in nanotechnology" today will be obsolete in ten years.

When I use *nanotechnology* in this book, I mean the original, atomically precise, sense of the word—the technology detailed in *Nanosystems*. (For brevity, I'll use *nanotech* as an adjective, as in "nanotech fabrication," instead of "nanotechnological.")

Where a term is needed to make the distinction, the best word seems to be *eutactic*. Eutactic means "well ordered" and has the same import in the context of nanotechnology as the phrases *atomically precise* or *low entropy*. Specifically, it means that in the system being talked about, there is a place for every atom, and every atom is in its place. Eutaxy is not necessarily nanoscale: a crystal is precisely ordered at the atomic scale, yet can be of macroscopic size. Turbine vanes in modern jet engines are grown as single crystals for strength at high temperatures. Quartz is eutactic but glass is not.

So what, then, is (eutactic) nanotechnology? It is a technology

Figure 1. Eutactic nanotechnology. In this schematic representation, a specific covalent bond is being formed on each of a series of identical molecular products. Courtesy of K. Eric Drexler (e-drexler.com).

that does not physically exist today, but can be straightforwardly analyzed, modeled, and simulated based on very standard, well-understood science and engineering. It involves building machines whose parts are of molecular size, but more important, of atomic precision: each atom and bond in the finished part is called for specifically in its design, just as the parts in the machinery of the cell are. In a mature nanotechnology, however, they will operate in a vacuum instead of salty water, and will be made of materials much stiffer and stronger than the proteins that cells use.

Given that, we can simply copy existing mechanical designs of assembly lines down to the molecular scale, and build as wide a variety of machines there as we now do at everyday scales. That's just a start, of course—many things will work differently and need to be redesigned. Some things will not work as well (centrifugal pumps, for example), but others that don't work well at macroscales will work at the nanoscale (electrostatic motors, for example).

Perhaps surprisingly, we can say more about such an ultimate technology than we can about all the different ad hoc techniques it will take to get from here to there. Imagine that you are Daniel Boone, leading a contentious band of bickering explorers westward. You climb a tall tree and can see the Cumberland Gap fifty miles away. All the explorers take off in different random, though generally westward, directions. Heaven only knows which path through the trees any one of them will take, but you still know that most of them will wind up going through the Gap.

We can see, in broad outline, a set of stages that we may go through, though not necessarily. There are many paths to the ultimate capabilities. However, this is one that seems reasonably likely, and includes some well-understood examples as landmarks.

STAGE I

Essentially what we have now—nanoscale science and technology—including the ability to image at the atomic scale with scanning probe microscopes, and a very limited ability to manipulate, that is, by pushing things around with the same scanning probes. A scanning probe is essentially like feeling something with a stick. Because you have a computer behind it, you can touch it in a very close grid of points and produce a picture. Also at Stage I are capabilities like chemistry. That is to say, you can produce an atomically precise product, like a molecule, but only by mixing chemicals or similar bulk processes. Thus you can have molecules whose atoms are precisely arranged, but the molecules themselves are dumped in random piles.

STAGE II

Now suppose we can take atomically precise parts, which we can form using chemistry or molecular biology. At the molecular scale there is a phenomenon called "self-assembly," which means that molecules will stick together in somewhat ordered ways. They act as if they were Lego blocks, but with little magnets instead of the pegs and holes. You could take a bucket full of such blocks, shake it vigorously, and expect the blocks to fall together in some semblance of structure. It was originally envisioned that designed protein molecules might be made to do this kind of thing; it has been done with tailored DNA molecules and also smaller molecules of a general chemical nature. Stage II is still in the lab, but some experiments have shown the ability to make larger structures with planned, complex patterns.

STAGE III

The next thing to do is to put the molecular Legos or Tinkertoys together, not just into pretty patterns but into functioning machines; in particular, machines that can put the Tinkertoys together themselves. This is the level of animal life: we are made of amino acid Tinkertoys, which our cellular machinery can put together to make more cellular machinery. At this stage we still require an externally created supply of the parts; we need proteins in our diet. Stage III will be the point where large-scale nanotech systems become feasible.

STAGE IV

The next stage involves the system being able to make the Tinkertoys from simple molecules, as plants do. This is the stage where we expect to see the beginning of a long-term trend of falling costs. Before this, the inputs to the process are expensive products of sophisticated chemistry and molecular biology; afterward, they are common and cheap.

STAGE V

The final stage is when the ability to make parts from simple molecules becomes general. In other words, instead of designing a part from the Tinkertoys, it's custom designed atom by atom. It's the difference between a machine made of actual Lego blocks and one made of custom-designed parts made with the same technology used to make Lego blocks. This is considerably more challenging technically. It is also the stage where the more remarkable capabilities of nanotechnology, as compared to life, for example, will appear.

All the work in nanotechnology, and today's nanoscale technologies, whatever its proximate goals, is leading in one general direction: toward the ability to build things smaller, with greater precision. Thus all the different paths ultimately lead to the same place: a technology where we can design things atom by atom, and build them as specified. What's more, we have an example of such a technology: it's biotechnology. The DNA blueprints and all the molecular machines inside the cell are built with atomic precision. So we know it's possible.

The science, as distinct from the engineering, of what happens at the atomic scale has been well established for half a century. It's the quantum mechanical wave equations of the same Erwin Schrödinger who wrote *What Is Life?* The equations tell you how electrons will arrange themselves to form atoms, and account for most chemical phenomena. In 1950, they were something of an ivory tower accomplishment because the techniques of the day couldn't solve them for most situations of interest. Nowadays, however, with modern computers, we can, at least enough to be extremely useful. So we can use this knowledge to design and simulate the machines that will be possible at the far confluence of the various paths that today's nanoscale science and technology are taking.

TECHNOLOGICAL REVOLUTIONS

Normal science, for example, often suppresses fundamental novelties because they are necessarily subversive of its basic commitments. . . . In these and other ways besides, normal science repeatedly goes astray. And when it does[,] then begin the extraordinary

investigations that lead the profession at last to a new set of commitments, a new basis for the practice of science. The extraordinary episodes in which that shift of professional commitments occurs are the ones known in this essay as scientific revolutions.
    —Thomas Kuhn, *The Structure of Scientific Revolutions*

Humanity has been through several technology-driven transformations before. The first two, significant fractions of a million years ago, happened so long ago they have shaped our biological evolution. They were stone tools and the techniques for using fire. A third major technology, agriculture, has been around long enough for some evolutionary adaptations to start, but they are not complete. On the other hand, agriculture has had a profound influence on the social structure and physical circumstances of humanity and was a necessary precursor to urban civilization. The various technologies involved in making clothing had a similarly profound effect, allowing humans to live in places like Europe and North America.[10]

Ships and wheeled carts made possible significant commerce. Reading and writing extended the spread of knowledge beyond the meager reach of the human memory. The scientific method itself was an invention, perhaps the most important one of the past millennium. The printing press takes first place among physical machines. Science and printing went hand in hand, enabling the industrial revolution and ushering in the modern world.

Nanotechnology has the potential for increasing our physical capabilities more than did the industrial revolution, expanding our ability to learn and communicate more than did the printing press, accelerating our ability to travel more than did the boat or the wheel, and enlarging the range of places we can live more than did clothing. It could induce greater biological changes in the human organism than the difference between humans and chimpanzees, indeed, greater than the difference between humans and horseshoe crabs. It is coming, possibly in the next decade, probably in the next twenty-five years, almost certainly in the twenty-first century.

CHAPTER 2

A HANDLE ON THE FUTURE
Can You Take These
Predictions Seriously?

There have been several attempts at setting up electronic mail within the federal post office administration, but all have failed, not for technical reasons, but because the post office is resistant to change, as it is locked into a bureaucratic system in which patronage and civil-service job security are far more important than efficiency. Given that history, it seems to me more likely that electronic mail will come about through private rather than governmental action. . . . At the present rate of development, electronic mail is probably no more than one or two decades away.
—Gerard O'Neill, *2081* (in 1981)

Things change. In the distant pretechnological past, Ecclesiastes' "preacher" could state that "there is nothing new under the sun," because the rate of change didn't allow one to see big differences in a single lifetime. That's no longer true.

By and large we predict the future by seconds or minutes, as in making a cup of coffee, or hours and days, as in buying tickets for a ball game. Foresight, as in knowing that you'll be asked for a ticket at the gate when you get to the stadium, consists largely of projecting into the future the situations and results you've seen in the past. The ability to manipulate this kind of knowledge is intelligence, and a reasonably comprehensive set of experiences to go with that ability makes it into common sense.

Predicting the far future in broad terms is thus problematical. We simply don't have the applicable experience. The best we can do is to fall back on broad generalizations that we hope will continue to hold true and models that we hope will continue to be valid. When I open the refrigerator to get cream for my coffee, I'm relying on a model that involves initial conditions from my observations last time I opened it and rules about what happens inside when the door is closed. (The cream spoils at a much slower rate than if I had left it out.)

Every general futurist gets the future wrong to a significant extent more than a couple of decades ahead. (I'm sure I'm no exception.) The most common failing, by far, is the simplest one: to assume the future will be like the present. Not surprising; that's how our basic future-predicting mechanism works, and it seems to require abandoning common sense, our store of experience, to predict anything else.

The point is important enough to repeat: things don't change that much or that fast in some ways, but in technology they do. In 1981, Gerard K. O'Neill, the Princeton physicist, wrote a book of futurism (entitled *2081*).[1] In his introduction he reviewed a collection of futurist writings from the previous century and noted that they tended to overpredict political change, but underpredict technological change. Arthur C. Clarke's classic work of prediction, *Profiles of the Future*, likewise begins with two chapters quoting hilariously myopic predictions by leading scientists regarding the impossibility of the electric light, railroad travel faster than fifteen miles per hour, heavier-than-air flying machines, and so forth.[2]

Historian of technology D. S. L. Cardwell writes:

> If we turn to contemporary speculation in order to gain some idea of men's expectations of the technology of their times, we find that in their predictions of the future, or rather their extrapolations of contemporary technological trends as they interpreted them, writers often made shrewd prophecies. Following the inventions of telegraphy and telephony, television could readily be imagined; air travel by heavier-than-air machines (usually powered by steam-engines and therefore boasting handsome funnels) was confidently predicted long before the Wright brothers' first flight. Even the atomic bomb was, it is claimed, forecast by H. G. Wells not very long after the beginnings of sub-atomic physics.
>
> It is, however, a truly remarkable fact that on the very brink of an economic-technological revolution unparalleled in history no

one foresaw the *universal* motor car and all that it was soon to imply. This failure on the part of informed and perceptive men to grasp the significance of what was going on under their very noses must make us suspicious of all attempts to forecast technological developments even one or two years ahead, much less ten or twenty.[3]

What fooled the prophets of 1900 was not the concept of the motorcar—that was well understood. It was the notion that the burgeoning productivity of the industrial revolution might carry through and provide the power of machines to a wide majority of the people. This was perfectly foreseeable: the numbers allowed for it, examples of various kinds were there. One of the major products of the earlier parts of the industrial revolution was cloth, which was made much cheaper and available to everyone. But it defied common sense that the average person could ever afford a complex, powerful, expensive machine not unlike a locomotive.

To do better, we have to tease apart the things that change and those that don't. The best set of unchanging facts we know is science: basic physics, chemistry, biology, and the like, and their handmaiden, mathematics. Another thing that doesn't change much is human nature, economics, and politics. One gets the feeling, on reading Aristotle's *Politics* and *Physics*, that he would have been flabbergasted by today's technology, but right at home with our politics.[4]

The things that change are sometimes random, in which case there's not much for us to say. But change often occurs within bounds or along predictable curves. Nobody can say who's going to have a baby next year, but population on the average tends to increase in a smooth exponential progression. It may or may not rain next July 4, but it will be warmer than January 4 (in North America, anyway). I don't know what specific inventions will be made next year, but the progression toward more complete control of physical phenomena is as old as the human race. Every year for the past century there have been inventions and discoveries that allowed us to see more clearly, manipulate more finely, and understand more deeply the physical world. This is not suddenly going to stop.

When we put that technological trend up against a model, we can get a reasonable prediction for future technology. The model in this case is basic science. Technology will get more and more capable, but it won't be able to do things that are impossible. Physics and chemistry will certainly accumulate more knowledge over the coming years. It's possible that there will be new phenomena discovered that will make some things we now believe impossible, possible; it's not so likely the other way around. The basic theory of what goes on at the human to the planetary scale—Newtonian physics with Maxwellian electromagnetics—has held up for well over a century, and the theory of what goes on at atomic scales, quantum mechanics, has been around for a lifetime. We need only imagine the relentless improvement of technological capabilities reaching the limits imposed by scientific law as we understand it, and we'll have a decent first cut at what a future technology could look like.

THE CURRENT TREND

Moore's Law is a rule of thumb regarding computer technology which, in one general formulation, states that the processing power per price of computers will increase by a factor of 1.5 every year.[5] This rule of thumb held true from 1950 through 2000. The factor of a billion improvement over the period is nearly unprecedented in technology. The end of Moore's Law has been predicted many times since the 1960s, but it's always been a false alarm. Once doped-silicon field-effect transistors (what's in your current computer) hit a size and speed limit, nanoelectronics can take over. It's already known how to build a molecular switching element, or rather, several ways are known. Just to prove the point, K. Eric Drexler designed mechanical switching elements that could be used to build a gigahertz computer the size of a bacterium.[6]

In other words, don't expect Moore's Law to end in the next two decades. In fact, if you project the Moore's Law curve, you'll find it as good a guide as any to the expected development curve of nanotechnology.

The key point here is that what's increasing so fast is processing power *per dollar*. That means that on the same budget, you can do 50 times as much computation as you could a decade ago, and 2,500 times as much as you could two decades ago

Among other effects, this explosion of processing power, along with the Internet, has made the computer a tool for science of a kind never seen before. It is, in a sense, a powered imagination. Science as we know it was based on the previous technology revolution in information: the printing press. The spread of knowledge it enabled, together with the precise imagining ability given by the calculus, produced the scientific revolution in the seventeenth and eighteenth centuries. That in turn led to the industrial revolution in the nineteenth and twentieth centuries.

The computer and Internet are the calculus and printing press of our day. Our new scientific revolution is going on even as we speak. The industrial revolution to follow—nanotechnology—is just beginning. If you extend the trend lines for Moore's Law a few decades, part sizes are expected to be molecular and the performance-price ratios become nearly astronomical. If you project the trend line for power-to-weight ratios of engines, which has held steady since 1712 going through several different technologies from steam to jet engines, we expect molecular power plants in the 2030–2050 timeframe.

The result of this will be essentially a reprise of the original industrial revolution, a great flowering of increased productivity and capabilities, and a concomitant decrease in costs. In general, we can expect the costs of high-tech manufactured items to follow a downward track as computers have. We can make such a prediction with relative confidence since we have the long-established trend and the scientific model that shows we have headroom for the trend to continue.

When a cat stalks through close spaces in near darkness, she avoids getting stuck in tight passages with her whiskers. They stick out the sides of her face and allow her to judge the width of openings by feel. Many of the designs and devices you'll learn about in the rest of this book aren't exactly predictions of future technology. They're cat's whiskers. I can show you, for example, the trend line for engines that indicates that we should have ones with more than a million horsepower per pound by 2050. But we can't know that physical law allows such a thing unless we come up with some kind of design and analyze it. By the time most of the gadgets we talk about here can actually be built, our specific designs will probably look pretty old-fashioned and clunky. But our designs feel out the headroom in physics to let us know that there's nothing to stop trends like Moore's Law.

Our designs are like Babbage's Difference Engine in some ways. Once computers got going, no one built them out of ratchets, shafts, and gears. But we could (and the Science Museum in London did in

1991, just to prove it).[7] We don't build them that way because we have something better. But Babbage could know with confidence, more than 150 years ago, that a computer at least as good as his design could be built, and indeed he was right. Where the whiskers fit, there will sooner or later come the cat.

THE SHAPE OF THINGS TO COME

Nanotechnology is specifically the technology we predict when the tide of technological progress washes against the shore of atomic physics (the quantum mechanics of electrons, with nuclei considered as unchangeable, primitive particles). Other technologies further in the future may well involve other shoals of knowledge, such as nuclear and subnuclear particle physics, and relativity. But these will come later and are beyond the scope of this book.

So what can we do with just atoms? Turning lead into gold is out—transmutation of elements involves changing the nuclei. Flash Gordon spaceships are out—nothing the size of a panel truck can hold enough energy in chemical form to travel from one planetary surface to another unassisted. Psychic powers? Nope, well outside the bounds of known physical law. Antigravity, as apparently used by Superman and the giant flying saucers in various science fiction movies? Sorry!

On the other hand, nanotechnology could turn sewage into breakfast. Transcontinental flying cars are a distinct possibility. Total, all-senses virtual reality, done by injecting signals into your sensory nerves or brain, is on its way. Towers not just miles high, but hundreds of miles high, are quite feasible. The countertop synthesizer or "matter printer" will revolutionize the way you acquire household objects and food.

Consider this: Your waste stream—garbage and sewage alike—is transformed into enriched soil by decay bacteria. Grass and trees grow on the soil. Cows eat the grass. The same atoms you discarded are rearranged into steak, milk, wood, and apples by natural molecular machinery. We can build molecular machinery to do the same thing. Coal into diamonds? Almost trivial by comparison.

Turning sick bodies into healthy ones or aging bodies into young ones is another transformation involving the rearrangement of atoms. These cases are quite a bit more complex than the previous ones, since we don't yet understand what actually needs to be done on the molecular level. But nanotechnology will first give us the tools to find out, and then the tools to do the job.

Will nanotechnology be able to build anything that's physically possible, as far as the arrangement of atoms is concerned? Probably not. First, for some arrangements that would seem to be physically stable if they did exist, there may be no physically possible way to get there from here.

A much larger class of structures are those that seem theoretically possible, but aren't in the practical world. For example, Ralph Merkle has pointed out that physics has no problem with a cube of diamond a meter on a side (i.e., a flawless three-and-a-half-ton diamond), which is a perfect crystal down to the atomic level. And in fact, nanotechnology could (and probably will) make such a diamond cube. In practice, though, it couldn't be perfect at the atomic level, because cosmic rays (highly energetic ionizing particles that are sleeting through our bodies all the time) would knock plenty of the atoms out of place.

Yet a larger class of things nanotechnology won't make are the ones requiring the development of new manufacturing techniques that, while quite feasible, don't have any useful application. Similarly, our current technology could make, in theory, many things for which the particular machines and techniques necessary will never be developed. (For example: a single-crystal pure silicon coffee mug, whose inner surface is lined with circuitry etched into the silicon just the way it is on a chip. That's clearly something current technology is capable of in theory, but it would require retooling a $10 billion fabrication plant to actually build.)

And of course, the biggest category of things nanotechnology won't make is the group of things that are possible, feasible, useful, and economical, but which we just aren't smart enough to think of.

WHEN?

I was gratified to be able to answer immediately: I said I didn't know.
—Mark Twain

It's possible that nanotechnology already exists as the result of a secret government effort like the Manhattan Project. Whether such a project could have succeeded by now depends on a whole raft of variables. It's getting close to the point that it would be irresponsible not to have such a project. However, it's not clear that such a secret military attainment of nanotech capabilities would make all that much difference to the course of commercial development. It would not accelerate development, since the military version would be kept secret. There probably wouldn't be much hindrance, either, since any significant activity of that nature would both arouse suspicions and hurt US economic competitiveness in the relatively near term.

So we can hope that secret early development happened, as a hole card against unexpectedly early development by malicious forces elsewhere in the world. But most likely, we can ignore the possibility without invalidating our analysis of mainstream development.

The road to nanotechnology, as Drexler described it in his *PNAS* paper,[8] and in much greater detail in *Nanosystems*,[9] was as a more or less straightforward extension to biotechnology. We already know the genetic code, that is, we can specify a sequence of DNA that will tell the ribosomes in a cell to create a protein with a specific sequence of amino acids. It's the sequence that causes a protein to fold up into a specific shape, and the shape that determines its function as a molecular machine.

Determining the shape that a protein will fold into is called, appropriately enough, the "protein folding problem." It's one of the holy grails of current molecular biology (part of the subfield known as proteomics). When Drexler's paper appeared in 1981, it was essentially an open problem. At this writing, good techniques are known for special cases, and plenty of progress is being made.

However, as Drexler pointed out in 1981,[10] there's a big difference between science and engineering. It's a distinction often lost on journalists, so when they need a pronouncement on the likelihood of some technological advance or another, they have a tendency to ask a scientist. This is often a mistake.

Science, historically, tends to lag behind engineering. People had been burning candles for millennia before Michael Faraday's famous lectures. The makers of wine and cheese had been using microbes for centuries before Louis Pasteur identified them. Steam engines had been in use for a century before Sadi Carnot elucidated the principles of thermodynamics. Celluloid and Bakelite were in use for decades before chemists finally understood that polymer molecules were long chains of their monomer constituents.

Both science and engineering involve the discovery of natural law and its application to the real world. The difference is that the scientist is trying to discover the most general law possible, applicable to any situation, while the engineer is trying to discover recipes to build machines that work in exactly one situation: when built as specified.

Imagine it's the 1600s and we are comparing Sir Isaac Newton with a military engineer of the day. Newton's law of gravity, one of the towering achievements of the human intellect, is applicable to all masses anywhere in the universe. The engineer is interested in a cannonball of exactly ten pounds traveling up to a mile from his gun. He can tell you how much powder to put in the gun and what elevation to give the barrel—because he's tried it. Newton's theory can't do that, because nobody knows how much energy the exploding powder produces, or how efficiently that energy is transferred to the cannonball, or how much much the ball is slowed by friction with the barrel or the atmosphere on its flight. The engineer doesn't know these things, either, but he does know how much powder to use to shoot a mile. The chemistry and aerodynamics (supersonic aerodynamics!) necessary to figure all this out from first principles didn't come along for centuries.

Similarly, as Drexler observed, you don't have to solve the protein folding problem in general to do the opposite: design a protein that will fold up a specific way—called de novo protein design.[11] You just use the sequences of folding you understand, and don't use the ones you don't. Sure enough, within a decade, it had been done, and by now it is a substantial field of its own.

The next step is to design proteins that don't just sit there, but do something. A protein that does something is called an enzyme. Progress is being made in this field as well, although we still can't call up a protein engineer and ask for a protein that will catalyze a given reaction.

Where does that get us? For an analogy, imagine that you are an explorer who has crashed on a hidden plateau like Sir Arthur Conan Doyle's *Lost World*. Instead of dinosaurs, however, what you find is a whole civilization of friendly, helpful people with an advanced technology. The catch is that their technology is based entirely on plastic. They have, let us say, nylon, polyethylene, and styrene. They can make any shape they like in these plastics and have a broad, highly capable technology based on them. They can make cloth and clothing. They print books. They have beautiful buildings of outsized

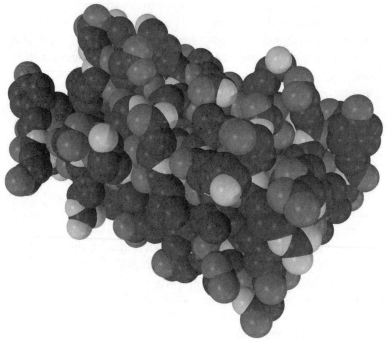

Figure 2. Alpha-4. The first intentionally designed protein, by DeGrado et al. in 1988.

Lego blocks. They have windmills, waterwheels, and even large but underpowered steam engines, which they use to drive stately boats slowly across their placid waterways.

But you'd like to get home, and to do that, you need an airplane. You can build many parts of an airplane from their materials, but you need metal for the engine: no engine of plastic has anywhere near the power-to-weight ratio you need. They have no metal—but the ground, you notice, is rich with metal ores, and there's plenty of clay. What you do next is a fun little exercise in imaginary engineering. Basically you do something like have your friends build plastic molds that you can use to build clay molds and crucibles with which you can make metal parts. Then you work your way up to a machine shop in which you can make the precision parts needed for an engine Depending on the specific ores available, the process could go lots of different ways, all of which would lead in the end to a metal-based metalworking technology.

The plastic technology is, of course, life. Biotechnology has a complete, self-sufficient manufacturing base, which is even self-reproducing. All we need to do to get to full-fledged nanotechnology is to bootstrap it to produce the kind of parts that can move it out of its current water-based, temperature-sensitive strength and power-poor limitations. This approach is called the bottom-up approach to nanotechnology.

THE VIEW FROM A HEIGHT

Another person who foresaw nanotechnology, in broad outline, was Richard Feynman, the second-most famous physicist of the twentieth century. In a now-celebrated talk given in 1959, he explained why "There's Plenty of Room at the Bottom." Much of the talk sounds quaint now, more than four decades later: Feynman foresaw things like the advantages of ultraminiturization in electronics for computers that we take for granted today. You have to remember that at the time computers with discrete transistors were the very latest thing, and most still used vacuum tubes.

Feynman had a different notion about how to proceed to a broadly capable technology at the molecular scale: start with the broadly capable technology we already have at the everyday scale, and make it smaller:

> Now comes the interesting question: How do we make such a tiny mechanism? . . . [I]n the atomic energy plants they have materials and machines that they can't handle directly because they have become radioactive. To unscrew nuts and put on bolts and so on, they have a set of master and slave hands, so that by operating a set of levers here, you control the "hands" there, and can turn them this way and that so you can handle things quite nicely. . . .
>
> Now, I want to build much the same device—a master-slave system which operates electrically. But I want the slaves to be made especially carefully by modern large-scale machinists so that they are one-fourth the scale of the "hands" that you ordinarily maneuver. So you have a scheme by which you can do things at one-quarter scale anyway—the little servo motors with little hands play with little nuts and bolts; they drill little holes; they are four times smaller. Aha! So I manufacture a quarter-size lathe; I manufacture

quarter-size tools; and I make, at the one-quarter scale, still another set of hands again relatively one-quarter size! This is one-sixteenth size, from my point of view. And after I finish doing this I wire directly from my large-scale system, through transformers perhaps, to the one-sixteenth-size servo motors. Thus I can now manipulate the one-sixteenth size hands.

Well, you get the principle from there on. It is rather a difficult program, but it is a possibility.[12]

This is called (surprise!) the top-down approach. Like life, it is autogenous. An autogenous technology does not necessarily mean a specific self-reproducing machine. It more generally means a set of manufacturing equipment that includes as a subset of its output everything that is necessary to make more of each kind of the equipment. No industrial base could survive without this property: some unmanufacturable machine would break, and things would grind to a halt.

Let's consider what the pure Feynman-style path to nanotechnology might look like. Essentially we would build an autogenous system in whatever macroscale technology we could, and then use it to rebuild itself iteratively, but on a smaller scale each time.

This is not as simple as it sounds, or it would have been done long ago. It's tough enough to have a milling machine produce objects as precise as its own parts, for example, much less improve them by a factor of four. Then one must have a design that is scale-invariant, or a way to change designs as scaling laws invalidate techniques. One of the advantages of this capability-first, size-second approach is that when scale-affected techniques do give out, you have a broadly capable system you can use to experiment with alternative techniques.

The system must be able to do handling and assembly as well as fabrication, since those operations cannot be done by a human operator as in a macroscopic machine shop. It must be able to "see" what it is doing, for the same reason. Feynman proposed using what are now called "waldoes" (after the Heinlein story).[13] Nowadays, robotics and teleoperation is somewhat more advanced, so we might imagine something more sophisticated—or not; whatever works.

The trick, of course, is to design robots as simply as possible, to reduce the complexity of the system. While we're at it, we need to pick techniques that can scale down as far as possible. An example is to build robots that operate by feel instead of vision, even though that may be slower and less efficient at the macroscale: at the nanoscale, vision won't work but touch does. Electrostatic motors work at the nanoscale but electromagnetic ones don't.

One difficulty for a machine, or a shop full of machines, that tries to build a copy of itself, is that all the design trade-offs that help you make precision parts tend to force you to build parts smaller than the manufacturing machine. One way around this involves a two-phase system which has big fabricators that make small parts, and assembly robots that put the parts together into fabricators and robots. (In a sense, that is how a standard machine shop works, with people as the robots, except the way people make other people isn't by putting parts together.) Such an architecture can be fairly scale-independent.

So how many steps would it take to get to molecular scale? The first thing to realize is that you have reached molecular scale not when your parts are the size of atoms, but when your tolerances are. A typical fine machining accuracy for most of the twentieth century was a ten-thousandth of an inch (2.5 microns); there are specialty shops that will do 1 micron. Atomic precision is about an angstrom (one-tenth of a nanometer), so you need to improve by a factor of ten thousand. If we can quarter the tolerance at each stage, as Feynman suggests, we are about seven steps away. How long would it take for each step, in a well-funded project? Maybe a year, maybe two.

Given current technology, though, you can skip way ahead. The electron beam machines currently used for nanolithography can cut shapes with a precision of about 2.5 nanometers. That's a thousand times better than the machine shop standard. With an e-beam machine, as they are called, you're only about a factor of twenty-five away from atomic precision: two factor-of-five steps or three factor-of-three steps. The problem is that there isn't a full machine technology at that scale. So it's a lot harder to take the shapes you've cut and put them together into a machine. Starting with a machine shop or nanolithography are the extremes of one set of choices—you can jump in at any scale in between. The smaller you start, the fewer steps to atomic scale, but the more limited your manipulation capabilities are initially.

Ultimately, the question of when devolves into the questions of how much effort is put into which pathways. If the billion dollars a year being spent by the National Nanotechnology Initiative were focused on molecular manufacturing in a well-directed way, we could almost certainly be at Stage III within ten years, and at Stage V in another ten. An Apollo Project–scale effort might cut that to five and five. But none of that is happening. The NNI's billion is being spread over a wide variety of science and technology, and the only thing it all has in common is that something about each piece can be measured in nanometers. That's just the way political funding works.

That's not to say the money is being wasted, by any means. It's going to good and valuable research in areas that have shown promise. The US government wastes billions each year, but this isn't where. It's just that a well-conceived, sharply focused effort might well succeed faster with less money.

WHERE?

There was an amazing flowering of science in France at the dawn of the nineteenth century. D. S. L. Cardwell writes, "During the years 1790–1825 France had more scientists and technologists of first rank than any other nation ever had over a comparable period of time."[14] We can mention Sadi Carnot, Antoine Lavoisier, Pierre-Simon Laplace, Joseph and Jacques-Étienne Montgolfier, Pierre Dulong, Alexis Petit, Jean-Baptiste Biot, Augustin Fresnel, Joseph Gay-Lussac, André Ampère, Félix Savart, Joseph Fourier, Gaspard-Gustave Coriolis, Augustin Cauchy, and Jean Lamarck—and these are just the ones whose names are attached to scientific laws and inventions that have survived to the present.

What's more, the advancement of science and technology was a well-funded national policy. Carnot was a graduate of the École Polytechnique, an institution that had no parallel in England or anywhere else at the time. England by contrast, far from supporting its scientists, allowed top-caliber people to be hounded for religious or other reasons. While the leading scientists of France were given seats at the École or other institutions and expected to teach the next generation of French scientists, Joseph Priestley, the discoverer of oxygen, was run out of England for his religious beliefs, and lived out his life in America.[15]

Besides the names, the textbooks and technical literature of France were notably superior. Charles Babbage, forefather of the computer and holder of the Lucasian chair at Cambridge (the professorship Newton had held), writes of searching for French calculus books and starting a society for the adoption of the Leibnizian notation over the cumbersome Newtonian one.[16]

It was not only an academic leadership. France was the acknowledged leader in most fields of actual technology. From the Jacquard loom, an early example of automated control, to the invention of the balloon and parachute, the French were ahead. Their roads, bridges, and cathedrals were better. They even built more advanced ships than the British at the time. French policy included public recognition and prizes for scientific and technical discoveries, and public funds were available for the development of new inventions.

So if you are a technological forecaster, what do you think happened next? What actually happened, of course, is that the industrial revolution occurred in Britain, not France. By 1850, Britain had railroads; Britain had steamships; Britain had the leading engine, machine tool, and textile industries in the world.

If you plot the amount of warfare going on all over the world from 1500 to 2000, you'll notice an odd fact. It holds a fairly constant average level throughout the period, with the exception of the nineteenth century. The level of warfare was consistently and significantly lower in the 1800s. This was the Pax Britannica: Britain held a global empire with a technology so much more advanced that it wasn't much challenged. It's not that the British were particularly peaceful: they began the century in the throes of the Napoleonic Wars, and ended it in the Boer War. Nor were they particularly more warlike than average, either. But they had the edge, so the world at large enjoyed a century of relative peace.

It is of more than academic interest to understand how this happened. The industrial revolution of nanotechnology has the potential to wreak a greater change on society than the one of steam. And it's clear that, as the players enter the stage for our rerun of this little historical drama, the United States is playing the role of France.

Historians have debated the issue endlessly and offer reams of theories. Those that ascribe some advantage to England per se, however, we can dismiss. Fast-forward to 1900 and we find British technological dominance in decline, and America on her way up. More likely is the analysis that notes the centralized structure of French science and technology. Paris was a brilliant center of learning, but it drew the best minds from the rest of the country. In some sense, the French system developed an ivory tower and insulated it from the practical problems on the ground. Thus, great strides were made in theoretical science; but practical techniques, which did not win the acclaim of the learned, received little attention. In England, almost perversely because of the official indifference to science, the best minds remained in contact with the real world.[17]

America today has a centralized science funding system and the effects can be seen in the political distribution of nanotechnology funding. One such effect—classic, well understood, and completely expected—is that most of the funding labeled for nanotechnology is channeled to people who are well-established scientists doing more or less the same things they always did. Lots of good and useful things will be done in these labs, and they will in fact ultimately lead to the kind of nanotechnology we are interested in here. But they won't do it as fast as the same amount of effort focused directly on nanomachine systems would.

In 1995 the number of doctorates in engineering earned in Asian universities exceeded those from US universities—and more than half of the engineering degrees in American universities were already being earned by students from Asia. Engineering doctorates from Asian universities continue to rise at more than 10% per year, but American engineering doctorates are actually declining.[18] It is all too easy to imagine a scenario in which India, China, Hong Kong, Singapore, Tiawan, South Korea, and Japan are competing to build nanomachines while the US funding agencies are still trying to soft-pedal them for fear of a genetically modified foods–style backlash.[19] Asian populations have more to gain, less to lose, and in many cases, less say in the matter, than Americans.

Americans have been in this kind of science slump before, and were startled awake by the sight of *Sputnik* serenely cruising over their heads. If we fall behind on nanotechnology, however, there is very little chance we could catch up. Imagine that the wake-up call is a breakthrough into Stage III nanotech—that someone has a mechanical version of a ribosome that can build copies of itself and other nanomachines out of molecular building blocks. Whoever has such a machine holds an enormous advantage over those who, like today's nanotech researchers, are still building every nanomechanism from scratch. Stage III systems could make nanoscale measurement equipment, test fixtures, and experimental prototypes of new machines—the technology base from which to build Stage IV systems.

The road to nanotechnology is like a five-mile race. You have to swim the first mile. Then you can run the second. At that point you find a bicycle. At the end of the third mile is a car, and at the end of the fourth is an airplane. The United States has more people in the water at the moment, but they're swimming in a broad group at random directions, mostly toward short-term goals, and not toward the next stage. If someone gets a significant lead, he'll be hard to catch.

THE FAILURE OF THE IMAGINATION

> When a distinguished but elderly scientist states that something is possible, he is almost certainly right. When he states that something is impossible, he is very probably wrong.
> —Arthur C. Clarke, *Profiles of the Future*

Scientists, in general, make lousy futurists. The job of a scientist is to establish facts by careful experimentation, documentation, and repeated testing until it is as verified as human inquiry can make it. If futurists waited for the kind of verification scientists require, they'd be historians.

In fact, a historian, particularly a historian of technology, makes a much better futurist than a scientist. Beyond the immediate horizon of what's in the lab today, the future is shaped by the same forces that shaped the past, and the past is one of our best guides for prediction.

Future technology is shaped by what people want. It's also shaped, of course, by what is possible. But a lot more is possible than people, particularly scientists, realize. The reason is that there are usually many different ways of getting a given result. A scientist can tell you, quite correctly, that a specific approach using a specific phenomenon won't work. But he can't know whether some other way, outside his narrow field of expertise, would reach the same goal.

An example very germane to nanotechnology is taking pictures of atoms. You can't use an ordinary microscope to take pictures of atoms because atoms are much smaller than the nature of light waves will allow you to focus. If you asked a microscopist, say, a century ago, whether it would ever be possible to take pictures of atoms, he would have said no, and been right by every rule of optics.

X-rays have the right wavelength but they have a couple of problems: there's nothing we can use to focus them and they don't bounce nicely off atoms but either go right through or shove them around.[20] By midcentury people were using x-rays to discover molecular structures, by an ingenious method involving crystalizing millions of molecules into a regular array and interpreting the scatter-shot diffraction patterns x-rays made off the crystals. But they would have told you, rightly, that nice photos of individual molecules, showing the rounded curves of the electronic surfaces, simply couldn't be taken.

Then in 1986, Gerd Binnig and Heinrich Rohrer won the Nobel Prize for doing just that.[21] But they didn't actually look at atoms; they felt them. The scanning tunneling microscope (STM), which they invented, uses a computer to control a needlepoint, atomically sharp, that "feels" the atoms by a very short-range electric interaction. The computer then puts the results together into a picture.

The STM came out of left field; nobody expected it. Once the principle was understood, a huge variety of similar scanning probes was invented, and pictures of atoms are commonplace today. Scanning probes are quite simple to build—they can be, and have been, built by high school students as science fair projects. But the point of interest here is that they use completely different physical principles than anyone would have guessed as the way to image atoms. A scientist trying to predict whether such a thing was possible would have gotten it wrong, ironically, by knowing too much about the way previous imaging technology worked.

Similarly, the laboratory scientist of today can tell you, quite rightly, that current fabrication technology can't make molecular machine parts out of diamond. But the historian of technology can point out three-hundred-year trends that point very strongly in that direction. The futurist can add some careful analyses and computer simulations and make more specific predictions. It is, after all, our profession.

In 1869, when Jules Verne wrote *Twenty Thousand Leagues under the Sea,* seagoing submarines were science fiction. In 1900, flying machines were science fiction. In 1950, space travel was science fiction. In 1963, a Robert Heinlein character "took his phone out of his pocket" while walking in a park to call a friend.[22] Today, an account of the future involving molecular manufacturing and nanomachines could be considered science fiction—but an account of the future without them is fantasy.

CHAPTER 3

CURRENT NANOTECHNOLOGY
Laboratory and Life—and How to
Get to the Real Thing from Here

> Most arguments against nanotechnology are arguments against life
> itself.
>
> —Marvin Minsky, in *Prospects in Nanotechnology*

Before diving into the nanotechnology of molecular manufacturing, let's have an overview of where we are now. This can be divided into two major parts: biology and current lab nanoscience. Then in each case we'll try to characterize the differences between what's current and what's coming.

NANOPARTICLES

As explained in the introduction, much of what you will see labeled as nanotechnology will be simply techniques that are extensions of current manufacturing and materials science, which happen to involve something smaller, shorter, or thinner than 100 nanometers. A classic example, and the most common one found in items on the market, is ultrafine powders.

When you stir sugar into your coffee, it dissolves. Sugar, like many substances, is made of molecules—groups of atoms held together by covalent bonds. The molecules themselves are held together by what are sometimes called van der Waals bonds and sometimes called nonbonded interactions—forces much weaker than the covalent bonds. When sugar dissolves, the molecules disconnect (each one remains intact), and float free, independently, within the coffee. Now pour in some milk. Milk is called a colloidal suspension, because there are things in the milk such as globules of fat that do not dissolve in water. However, the globules are small enough that they don't just sit on top in a layer like oil would, but remain dispersed, acting a bit like supermolecules.

Particles up to the size of a bacterium or so, a billion times the volume of a sugar molecule, can form suspensions like this. Flour is ground to this size at its finest. The reason flour is ground so fine is that it's useful in many processes to have a very even mixture of something that doesn't technically dissolve. The finer you can get your powder, the more even your mixture and the better properties it has for many applications.

The rubber used in tennis balls, for example, is mixed with clay before being "cooked" to give it the proper weight and consistency. If the clay can be divided into nanometer-scale instead of micron-scale particles, the resulting rubber is more resistant to air leaking through it, just as sand makes a better dam than gravel.

Many ultrafine powder applications are now being marketed as nanotechnology, and typically they offer increased stain resistance, easier cleaning, and similar improvements over previous versions where the powders were "merely" microscopically fine.

A recent product advertised as nanotechnology is Demron, the radiation shield material. It's a polymer composite of polyurethane and polyvinyl chloride that is mixed with a variety of various ultrafine powders. The powders are of substances that include heavy atoms, and they act in a concerted way to absorb and scatter radiation of various wavelengths. The overall result is a thin, rubberlike sheet that stops x-rays and the like, which no one would have believed possible a decade ago.[1]

Demron is an ingenious invention and a true advance in technology. The size of the particles essentially allows them to be spread more evenly and to force each x-ray to contend with more different substances in a given space. The set of different substances is the key; and then the particles' size allows the material to be wearably thin.

Coatings can have different properties at the nanoscale as well. Titanium dioxide is a brilliantly white substance that is used as a pigment in paints. But a layer of titanium dioxide only 15 nanometers thick is transparent, and can be applied to glass to make it self-cleaning. It has a photocatalytic effect that helps sunlight break down dirt, and a hydrophilic effect that causes water to form even films that carry the dirt away.[2] So a rain or spray from a hose will get the glass as clean as if you'd washed it with detergent.

CARBON NANOTUBES

One of the structures most commonly identified with nanotechnology in the popular press is the carbon nanotube, or "buckytube." The buckytube is essentially a strip of graphite rolled up into a cylinder. Graphite is amazingly strong—it's been used for years in pricey high-performance items like fighter jets and golf club shafts—but buckytubes could be stronger because, being seamless cylinders, there's no edge where a tear can start.

Buckytubes are big news in the nanoworld because they can be made today. You can buy them in bottles. The synthesis process is still a bulk technology; you can't specify what kind of nanotube you'll get, and you typically get a mixture of different kinds. The kinds vary by diameter and chirality. If you imagine a nanotube as being formed from a strip of graphite cut out of a graphite sheet, chirality means the angle you cut the strip with respect to the sheet's grain. (Chirality has a different meaning in ordinary chemistry.)

Nanotubes, as currently made, are marvelous for research. Lab scientists don't mind picking out ones with specific properties using scanning probes, and their properties are so varied that they are a wide-open field for research. The electrical properties in particular vary widely; some nanotubes are conductors, some semiconductors. Transistors and diodes have been made from nanotubes.

The other main property of nanotubes in their natural state is that they're slippery. Often, nanotubes form in a nested configuration, sometimes called MWCNTs, or Multiwalled Carbon Nanotubes. Inner and outer tubes slide and rotate easily against one another, like the wire and casing of a bicycle brake cable. This could be useful in many applications, but the slipperiness makes it difficult to use the nanotubes' strength in a composite material. However, special additives have been developed to help the matrix (the glue holding the fibers together in a composite) "grab" the tubes better.[3]

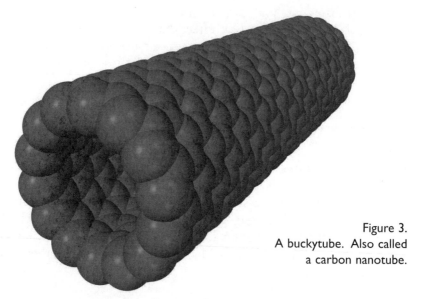

Figure 3.
A buckytube. Also called
a carbon nanotube.

MICRO AND NANOLITHOGRAPHY

Another technology that's been called nano is the lithography used to make electronic microchips. The size of transistors and width of wires on a modern microprocessor is typically under 100 nanometers today. The basic process is not unlike, in principle, painting using a sprayer and a stencil. Other parts of the process are like developing a photo where light makes a chemical change to parts of the chip. The masks for these can be large and are focused down on to the chip with lenses. Once the chemical changes are made, chemical processes are used to construct or modify layers of material in the patterns thus laid out. These techniques are good for making millions of copies of a working device at these scales (and underlie the computer revolution of the past few decades).

These lithography techniques can also be used to make machines with moving parts at that scale. These are larger, and more to the point, much less precise, than the nanomachines we talk about in the rest of the book. They are generally still called micromachines. Because of the process by which they are made, they tend to have a very two-dimensional character. Imagine trying to build a machine out of shapes that you cut from foam-core craft board. You can glue them together, if you like. The rub is that every one has to be lying flat.

The other rub is that they won't rub. Micromachinery has a phenomenon called "stiction," which is a combination of being sticky and having high friction. The stickiness is due to van der Waals attraction and the friction is due to the fact that the parts are not atomically smooth. This makes it difficult to make bearings, sliders, and so forth, which severely constrains the kinds of machines you can make.

Even so, there has been quite a bit of progress in micromachines. For example, Zyvex, a Texas nanotech start-up, uses the lithographic techniques to build manipulators that can be used to pick up and assemble other micromachines.[4]

The next step beyond microlithography is electron-beam nanolithography. Lithography with light is limited to micron sizes because the wavelength of light won't let you focus it any more finely. You can get some slack by using shorter and shorter wavelengths of light, up into the ultraviolet, but after that you're trying to use x-rays, which are difficult to handle. It's easier to handle electrons, which can have a nice short wavelength but still be manipulated by magnetic fields. Thus electron microscopes can see much finer details than optical ones.

Electron beams of sufficient power can carve material into shapes as well. They are used, for example, to make the stencils for ordinary lithography of microchips. Current top-of-the-line e-beam machines can cut shapes to about a ten-atom precision. The machines, of course, are expensive, and only make one part at a time, so commercial products can't be made this way. Even so, such parts are within shouting distance of the atomic scale if ways can be developed to manipulate them.

NANOMANIPULATION

The hump to be gotten over in manipulation is that third dimension. Manipulating atoms and molecules in two dimensions has been done routinely for more than a decade. A scanning probe microscope is essentially a hunk of quartz or other piezoelectric material, that is, one that expands or contracts slightly when a voltage is applied to it. We can carefully control voltages across our hunk and cause the far end of it to move, and we can control the movement with much better precision than the width of a single atom.

A number of techniques will produce a point that is atomically sharp. With skill, simply cutting a thin metal foil at an angle with scissors can work. Take such a sharp point and mount it on your hunk of piezoelectric material, and you have a tool that can be used to "see" or move atoms and molecules. The molecules stick to a surface by van der Waals attraction and can be pushed around with a lot of care and effort.

Imagine you have a plate full of peas that have been coated with honey. Put them on the floor, run upstairs, cut a hole in the floor above, and reach through with a fishing pole. You can now manipulate the peas about as well as we can manipulate atoms with a scanning probe. The pole is sticky, too, so you can drag and stack them as well as just pushing. But it's a tricky, time-consuming process. You could readily imagine being able to make two-dimensional patterns with your peas, but building three-dimensional machines, as you can imagine, is quite difficult.

On the other hand, if you had a plate of sticky machine parts you had to assemble into a robot arm that would then do all the work for you, you might be willing to spend a great deal of effort and keep at it through numerous false starts. Especially if it were worth a billion dollars to get the first one working.

Another thing you could do with your fishing pole is attach a quill pen and put an inkwell next to the plate. Then you could write, draw pictures, and so forth. If you had thick ink that dried hard, you might even be able to build up a structure of some kind. When you do this with a scanning probe, it's called "dip-pen nanolithography." You can see that the term "lithography" is getting about as stretched as "nanotechnology" itself. It's come to mean any kind of printing or writing process at all, instead of the original meaning of a specific contact printing method.[5]

A process that's more like the original lithography is sometimes called soft nanolithography. It would be more descriptive to call it "nano rubber stamps." Take a block of material with a raised design on one side, dip it in ink, and touch it to a surface. You get a print of the design. This works even if the stamp is a few nanometers in size.

Molding—melting something, pouring it into a mold, and letting it harden into the shape thus formed—also works down to the nanoscale.

MOLECULAR BEAM EPITAXY

If you have a CD or DVD player, there's a good chance you own a device that is atomically precise—in one dimension, that is. Many of the solid-state lasers used to read the disks are made by molecular beam epitaxy. This is a technique that involves growing an object by exposing it to a very diffuse gas of atoms, made by vaporizing small amounts of some (usually solid) material in a vacuum. This technique can achieve remarkably fine control over the thickness of the layers thus produced, indeed controlled, to within an atom of thickness. To get a desired sequence of layers, you expose your growing object to a series of different gases. A wide variety of devices, solid state with no moving parts, like a transistor, can be made using the quantum properties of the thinness of the layers.[6] Lasers, unusual kinds of transistors, Fabry-Perot etalons (like a transistor, but using light), and the like can be made.

NANOELECTRONICS

The most advanced capabilities seen in labs today are in the area of nanoelectronics. A wide variety of molecular-scale transistor-like switches has been discovered, and wires and memory elements fabricated using different phenomena. What's more, the path to making these into useful devices may well be shorter than one involving machines with moving parts. Electronics is the main application of microtechnology, for the same reason: microchips don't have to have moving parts.

The short way to nanoelectronics is by self-assembly. For example, you can synthesize your switches and wires chemically—after all, they're just fancy molecules—and chemically attach them to pieces of DNA. The DNA will have been carefully arranged so that when it self-assembles, it will drag the molecular electronic devices together into a working circuit. This will probably have been done in a lab by the time you read this, but it probably won't have been reduced to efficient enough practice to be commercially viable yet.

NANOMEMORIES

Another near-term product that seems likely is one that uses scanning probe techniques to read and write data at molecular densities. If you've ever built a crystal radio (from scratch, not a kit), you could probably build a scanning probe—and you will also have some understanding of how finicky and tricky they are to use.

Scanning probes were, after all, invented in IBM's labs. For almost two decades now, they have been capable of reading information that was written at atomic resolution. At a byte per square nanometer, this entire book would make a bathmat for a bacterium. The trick is in writing, reliably and quickly, and in reducing the scanning probe from a lab instrument for use by a skilled technician, to a completely automatic gadget you can put in a box and forget. That improvement will come about, in the main, with the ability to manufacture the scanning probes more precisely, and in quantity.

NANOBATTERIES

By far the slowest-moving part of the technology associated with computers is batteries. Batteries have improved some over the years, but everything else from processors to displays has improved so much faster that batteries seem to be moving backward. This will not change greatly until we have fuel cells of battery size and weight, but current-day nanoscale technology can offer a spate of modest improvements until then. For example, the amount of power a battery can produce is limited by the amount of surface interface between its electrodes and electrolytes. With a nanostructured surface, you can have a lot more area in a smaller space and with less weight.

For example, a really neat trick involves nanograss, a structure that consists of an array of silicon nanorods sticking straight up from a surface like grass from the ground. (They're all the same height and in regular rows and columns, so the fakir's bed of nails might be a more accurate image.) Droplets of electrolyte sit up on the tips and don't touch the sides, under normal circumstances. This cuts the actual interface area down to virtually nil, which reduces leakage and battery rundown to practically zero when it's not in use. When it is in use, a voltage or other stimulus can be used to force the droplet down between the spikes, increasing the surface area of contact enormously. And because the nanograss battery is silicon based, it is compatible to be fabricated as part of an electronic chip, producing self-powered chips for non-power-intensive applications.

Fuel cells are another area where current nanoscale technology could very well enable a significant advance. Quite a few different technologies are being looked at today. The basic idea is to find some way to do reverse electrolysis. You may have seen or done the experiment where you put two wires into water and run a current through it. Hydrogen bubbles up from one wire and oxygen bubbles up from the other—direct conversion from water to its constituent elements using current. The opposite is also possible: direct conversion of hydrogen and oxygen to water, producing current. This is what happens in a proton-exchange membrane (PEM) fuel cell.

PEM fuel cells are a current technology; you can buy them in a variety of sizes, ranging from power backup for your house to micro-sized units for your laptop. They are costly, as yet, compared to other power sources. The quip in the industry is, "Buy a micro fuel cell, get a laptop free."

Advances in nanoscale technology, such as carbon nanotubes for electrodes, might well be involved in the continuing improvement of fuel cells. At the same time, although I haven't heard of such a project, I wouldn't be surprised to discover a biotech-based project to harness the sugar-digestion machinery of the mitochondrion to run a fuel cell. Whatever works!

SELF-ASSEMBLY

Another way to put things together at the nanoscale is what's called self-assembly, or Brownian assembly. This is how the molecular machines inside the cell get put together. The trick here is that the parts to be put together have to be atomically precise (or nearly so). Then you dissolve them in some fluid, often water, and let them float around. When the ones that fit properly come together by chance, the stickiness (van der Waals and some other forms) causes them to stay together. Take two bricks and break each one in half. Apply a thin layer of glue to the all the broken surfaces. You'll find that two matching halves stick together much more strongly than two nonmatching halves.

How do we get atomically precise parts? Currently, we have to use chemistry or molecular biology. For example, if we can make a molecule of DNA, by whatever means, we can use some of the cell's machinery to duplicate it in great quantities. It turns out that DNA is a good molecule for doing self-assembly. Scientists, particularly Nadrian Seeman and his group at NYU, have been building some fairly complex structures, and even some moving machines out of DNA.

Their latest effort, as this book is written, is a pair of walking legs. The legs don't balance or anything like that. What the group has constructed is a pair of somewhat floppy legs and a DNA footpath. The "feet" stick to successive "footprints" in the path, unsticking from the last and going to the next by infusing the whole business with a series of different chemicals that alter the properties in just the right way. This makes for a cumbersome machine, requiring rinsing and mixing for each little controlled motion, but if you can get a machine that works, you can use it to build a faster one next.

FROM NANOSCALE TECHNOLOGY
TO NANOTECHNOLOGY

Currently, the applications of nanotechnology in products you can buy are things like the nanoscale powders that produce new (and useful) materials properties. There's some advance in catalysts and chemical analysis, mostly in specialized applications. Batteries, fuel cells, and data storage may well soon see some benefit from today's nanoscale science. But to get to the really impressive capabilities of nanomanufacturing, you have to be able to build nanomachines. Right now in the labs, a handful of experiments is addressing prototypes of individual parts, such as bearings, that will go into the nanomachines (see chapter 5). Once we do get all the parts, we still have to figure out how to put them together.

Imagine a set of parts you have to put together to build a machine. They have to go to a specific set of positions in a specific sequence. Suppose you could have each one made with a semiflexible rod connecting it to a sort of luggage tag that had a complex surface on both sides. The bottom surface would match the top surface of the tag of the previous part, and its top would match the bottom of the next.

The tags would self-assemble into a stack in only one sequence, and the rods would be such as to force the parts into the appropriate configuration, in the right order, as the tags were stacked in that sequence. You could complete the assembly process by dissolving away the rods.

The big difference between today's protonanotechnology and the full-fledged, mature technology that will forge the shape of our future is simple. Current practice involves factories that make nanoscale products. Mature nanotechnology will involve nanoscale factories: working machines, atom-for-atom precise, that make things.

LIFE: NATURE'S NANOTECHNOLOGY

Of course, the world is crawling with nanomachines already assembled, working, and making things. Each living cell is an amazing factory jammed with molecular machines. Most of these machines are made of protein, that is, long chains of amino acid molecules that are tangled up into a compact shape like a bead necklace rolled into a ball.

It's the specific shape of these protein molecules that gives them their function. Often a specific machine, or enzyme, is formed of more than one actual molecule, and the shape (and distribution of electric charge) of the molecules causes them to stick together properly.

This sticking together properly is the key to how everything in the cell works. A cell has a number of different compartments to keep various kinds of molecules separate, but nowhere near the total number of kinds of molecules. So in general, to make a machine composed of molecule A and molecule B, the machinery simply makes an A and a B and lets them float around until they bump into one another and stick. This floating and sticking process is called self-assembly, when you are talking about putting the parts of a machine together.

Once the protein machine is put together, it performs some function, for example, as an enzyme. An enzyme takes one or more input molecules and causes a chemical reaction to occur, producing one or more output molecules. The protein pocket that snugly fits the input molecule(s) holds it in such a way that the reaction is favored; that makes it a catalyst for that reaction. These reactions can be anything from building up amino acid molecules from smaller ones, to oxidizing sugar for energy, to breaking down old proteins for spare parts.

If the reaction "runs downhill," producing more energy than it consumes (such as oxidizing sugar), the enzyme can do it without any help. If it runs uphill, such as in building big molecules, the enzyme needs to get energy from somewhere. Typically, the energy comes from special molecules such as ATP that serve as fuel. The reaction then is more complex, involving breaking down the ATP at the same time as doing the other, desired, reaction. Mechanisms like this allow the enzymes to add energy to their products. But where does the ATP come from?

ATP (adenosine TRI-phosphate) is formed from, and broken down to, ADP (adenosine DI-phosphate) and an extra phosphate—the parts being used over and over to transport energy from one reaction to another. The energy comes from the oxidation of sugar and fat in the mitochondria in higher forms of life. (Bacteria have a different way of getting energy from sugar without oxygen, fermenting rather than burning it.) The mitochondrion is a double-walled compartment in a cell. The oxidation goes on inside the inner wall, involving a complicated cycle of reactions that ultimately winds up creating a voltage difference across the inner wall. It's similar in principle, though wildly different in detail, to what happens in a PEM (proton-exchange membrane) fuel cell.

The inner wall of the mitochondrion is covered with what looks like, in a micrograph, an aerial photo of a forest. The trees are molecular machines called ATP synthase. (Decoding the biologyspeak, that means "machine that makes ATP.") The ATP synthase machine has three main parts: an electric motor, a shaft, and a mill that takes ADP molecules and pops the extra phosphate on by mechanical force.

Most of the molecular machines in the cell are not this obviously mechanical, but ATP synthase shows that they could be. It demonstrates electric motors: quite a few motors are found in the cell, several of which are rotary. (Note that intracellular motors typically run on proton rather than electron transfer, but that's a technicality.) It shows a shaft connected to a motor, transmitting mechanical torque and power. Most important, it demonstrates a process that causes a specific "uphill" chemical reaction pushed by mechanical force.

One ATP synthase machine can produce something like one hundred ATP molecules per second.

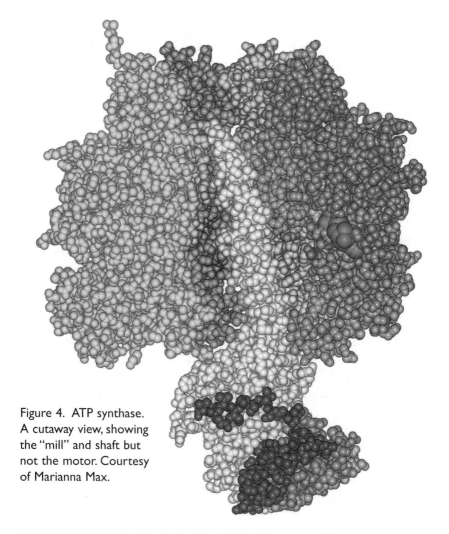

Figure 4. ATP synthase. A cutaway view, showing the "mill" and shaft but not the motor. Courtesy of Marianna Max.

THE MOTHER OF ALL MACHINES

Let's look at one other machine, out of the thousands in the cell, in some detail. It's the ribosome, and it is the key to all the molecular machinery in the cell, because it's the one that makes proteins.

A protein is a wadded-up string of amino acid molecules. Twenty different kinds of them occur naturally, and each has different properties. Some are stiff and some are flexible. Some are large and some are small. Some are charged and attracted to water; some are oily and like to bunch up away from the water. The shape a protein molecule ultimately takes depends on the sequence of types of amino acids in the protein. Thus, to make a specific molecule, you have to specify a particular sequence, and then build a string of amino acids that meets your specifications.

Protein sequences are stored in the DNA of the cell. DNA is a double string of just four kinds of submolecules (called nucleotides), so it takes three of them to refer to one amino acid type. The nucleotides are commonly denoted A, C, G, and T. So if we had a piece of DNA that read GCATGCAACCCA, we would parse it GCA TGC AAC CCA and it would mean (a tiny snip of) a protein consisting of alanine, cysteine, asparagine, and proline in that order. The correspondence between the nucleotide triplets and amino acids is called the genetic code.[7]

To make a protein, the information permanently stored in the DNA is copied off into a "work order" as a molecule of RNA, which is similar to one of the two strands of DNA. The RNA is then "read" by the ribosome to produce the protein. The work order RNA is called messenger RNA, or mRNA for short. Another object made of RNA, this one called "transfer RNA," or tRNA, acts as an amino acid holder, and thus comes in twenty different forms. Each form corresponds to one of the amino acids, and in fact, each one gets attached to a molecule of the appropriate amino acid by a special enzyme made for that specific purpose. It has the acid (e.g., alanine) at one end and a triplet (e.g., CGT) complementary to the acid's code on the other. The tRNAs are not just amino acid molecule holders; they are *labeled* amino acid holders.

The ribosome is a two-part machine, each part consisting of several protein molecules and one or two RNA molecules (rRNA, if you must know) that form their backbone. The two parts come together like the halves of a split mold and form tRNA-shaped cavities when they do. In operation, the ribosome holds a strand of mRNA along one edge, so that one triplet forms the end of a cavity. This causes the cavity to fit just the tRNA that holds the amino acid the triplet specifies. And in the cell, if there's an object, and a cavity that fits it, pretty soon the object will be found sitting snugly in the cavity. That's the magic of self-assembly.

The two parts of the ribosome reciprocate back and forth against each other, not unlike the slide and frame of an automatic pistol.

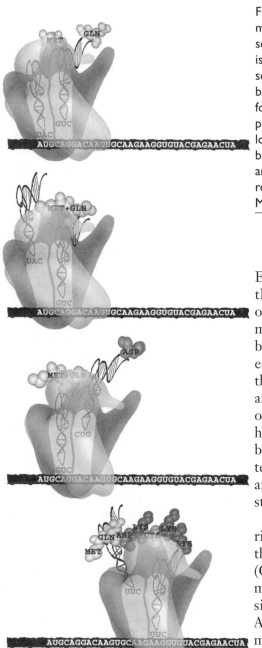

Figure 5. The ribosome makes a protein. In this schematic series, the ribosome is shown assembling the first seven amino acid subunits of brazzein, a sweet protein found in berries. The finished protein will be fifty-four units long. The dark stripe at the bottom represents mRNA, and the wireframe holders represent tRNA. Courtesy of Marianna Max.

Each stroke jacks another tRNA through the series of chambers, advances the mRNA "tape" one triplet, bonds the amino acid molecule of the latest tRNA to that of the previous one, and unbonds the previous one from its holder. The holders are popped out to be reloaded, and the protein "chain" grows by one amino acid "link" at each step.

Each stroke of the ribosome is powered by the breakdown of a GTP (Guanosine TriPhosphate) molecule into GDP, very similar to the ATP-to-ADP process that powers most of the rest of the cell's activities.

Ribosomes in human

cells can add two amino acids per second to the protein they're making, and in bacteria up to twenty.

FROM BIOLOGY TO NANOTECHNOLOGY

The molecular machines in the cell have many similarities with the ones that will form the core of nanotechnology. First and most important, they build atomically precise products by controlling specific bond-forming and bond-breaking reactions. They are—in instances like the ribosome—digitally controlled; that is, they receive instructions encoded as a sequence of discrete symbols. They can use electric motors, turning shafts and reciprocating advance mechanisms.

The synthesis process in the cell has parallels to the nanotech ones that have been designed to date. Feedstock molecules are broken down in a sequence of specific reactions. The moieties (parts of molecules) that are going to go into a product are put on holders, again by a sequence of specific reactions. Then a mechanism takes them one at a time and attaches them to a growing product.

But there are big differences, too. Nanotech machines will typically be filled with vacuum, not water. They will be able to use more reactive molecules, even ones with dangling bonds, in their processes. A molecule with a dangling bond is called a free radical and causes damage to the other molecule it bumps into. In a nanomachine, molecular parts will be carried on specific paths and thus not allowed to bump into things that they could disrupt. Moving along specific paths, on conveyor belts or in robot grippers, means that the parts will go faster and get to where they are going with more regularity. Thus a nanotech machine performing a specific reaction on a stream of incoming molecules should be able to do millions per second, instead of ATP synthase's one hundred or so per second.

Rather than distributing energy in one-shot cartridges (the ATP molecules), nanomachine systems will tend to do it with wires, shafts, cables, and pushrods. Rather than using the intricate control processes of the cell, involving chemical messengers that are still not completely understood, nanomachines typically use cams, linkages, relays, and computers.

The parts-assembly mechanism in a nanomachine would begin making a part because a computer program told it to, not because mRNA happened to run into an idle ribosome. The mechanism itself would look more like a robot arm. The arm would take molecule holders from conveyor belts, rather than waiting for one that fits to float by.

A big difference is that the nanomachine will build a three-dimensional part in three dimensions, rather than spinning a chain of beads that fold into a shape. This means that nanomachine parts could be much more rigid than proteins, and could be made with many fewer building-block types. The robotic arm might need five or fewer incoming conveyor belts of holders, instead of twenty. It would have many more choices of where to attach the new moiety, compared to the ribosome that can only tack it to the end of the string.

Stiffer, more rigid parts not immersed in water means that nanomachines must operate in a wider range of temperatures than living cells (below freezing, for example). Operating at higher rates of speed and using higher energy reactions are just the normal differences we expect between machines and living things. With nanomachines, though, we will get to keep many of the things about life that are due to its basis in molecular machinery: complex, efficient metabolism; powerful yet silent motion; intricate structure; self-repair; growth—in short, all those qualities that make life a wonderful and seemingly mysterious thing.

CHAPTER 4

DESIGNING AND ANALYZING NANOMACHINES
Theory, Current Tools, and the Future

O ne of the major reasons we can talk with certainty now about the kinds of machines we will be able to build in the future is that we can design and simulate them now. A wide range of computer programs for use by chemists encodes a great deal of the knowledge they have about how atoms and molecules behave.

THE WEIRD QUANTUM WORLD

The most basic theory relevant to chemistry is called quantum electrodynamics, or QED. In simple terms, QED tells us how photons and electrically charged particles, like electrons or atomic nuclei, behave. Physicists, of course, are interested in deeper stuff like how the nucleus behaves internally, but we can get along just fine without that.

In fact, we can get along fine without a lot of QED itself. This is good, because calculations using it are quite difficult. It would be a bit like using the theory of relativity to calculate where a baseball is going. There is no practical difference between the results of the much simpler calculation using Isaac Newton's laws of motion and the more difficult relativistic one. In the same way, for purposes of chemical interactions, we can almost always get away with using the Schrödinger equation. This still captures most of the quantum effects.

The quantum world is strange and hard to understand. Our

instincts about the way things work are formed in the larger world, where objects exist, and waves have to be waves in something, like sound in air, or at the surface of water. In the quantum world, both of these descriptions apply to some degree to everything. Something can seem to be an object, like an electron, but can act as if it were in many places, or moving in many different directions, or had many different energies at once. This odd property is called superposition of states and is the thing that will let quantum computers be able to calculate many different things at the same time, thus being vastly more powerful than ordinary computers.

By the same token, when you take an ordinary computer and try to simulate something in the quantum world, the calculations can easily get out of hand. You have to keep track of all the superposed states and the interactions between all the particles. The amount of work blows up exponentially with the number of particles. Therefore even when doing the quantum mechanics of a chemical interaction, the analysis is limited to a relatively small number of atoms, and something called the Born-Oppenheimer approximation is used. This approximation treats the nuclei as if they were ordinary objects and calculates the electrons quantum mechanically around them. Nuclei, it turns out, are heavy enough to act like ordinary objects most of the time.

Even so, a study such as that done by Robert Freitas and Ralph Merkle of the reactions involved in depositing carbon atoms to build up diamond machine parts required 50,000 CPU-hours of calculation.[1]

We know that the Born-Oppenheimer approximation doesn't always hold. Physicists can do experiments that cause superpositions with entire atoms being in two places at once, or meld together in overlapping states called Bose-Einstein condensates. But this is where the distinction between science and engineering is important. We can design machines that don't do those experiments, and only do the things we can analyze (relatively) straightforwardly instead. We know that atoms act like objects under most normal conditions—we can image them with scanning probes, push them around, put them somewhere, and find them still there when we look back.

MOLECULAR MECHANICS

If you're an automotive engineer, you need to analyze the chemistry in the fuel-air explosions in the cylinders of your motor, but you don't need to simulate the chemistry inside your structural steel to know that it remains steel from moment to moment. You use a higher-level approximation that involves weight, tensile strength, coefficients of elasticity, and so forth. The higher-level approximation that is appropriate for the parts of nanomachines where chemistry isn't being done is called molecular mechanics.[2]

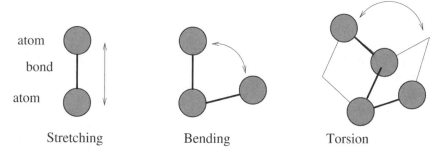

atom

bond

atom

Stretching   Bending   Torsion

Figure 6. Molecular mechanics. The forces along and between covalent bonds that are accounted for in a typical molecular mechanics force field.

Molecular mechanics treats atoms as objects and condenses the descriptions of what the electrons do around and between them into a number of simpler forces. First is the van der Waals force, which is a small attraction between atoms up to a point, and a larger force resisting their being pushed together after that. The van der Waals force makes atoms act like objects, somewhat soft and springy, but with a definite size.

Next are the ionic forces, which occur when atoms gain or lose electrons (or part of one, in a molecule). The resulting electric charge causes a longer-range attraction or repulsion between atoms. For example, ordinary table salt is a crystal of sodium and chlorine ions in which the sodiums have lost electrons and the chlorines have gained them. (*Ion* simply means an atom that has lost or gained one or more electrons.) The electrostatic forces that result make the atoms form a very strictly alternating pattern in the crystal. In the water molecule,

H_2O, the oxygen atom pulls the electron from each hydrogen in close, so that the hydrogen's nucleus (a single proton) is left sticking out a little. The result is that the hydrogens act positively charged and the oxygen acts negatively charged, though not as strongly as if the electron had been pulled all the way off. Thus water molecules stick to each other, hydrogen to oxygen, and stick to the ions of salt, which is why salt dissolves so well in water.

Finally comes the force associated with a covalent bond. This is where, instead of one atom grabbing the electron and the other letting go, they both hold on more or less equally. (Actually, the electrons are shared in pairs.) A covalent bond acts as if there were a strong spring between the atoms, holding them much closer together than the van der Waals force would let them, because they actually overlap where the shared electrons are.

Covalent bonds are much stronger than ionic forces, and ionic forces in turn are much stronger than van der Waals ones. Diamond, for instance, is a crystal held together with covalent bonds, and salt is one held together ionically. For a van der Waals crystal, you'd have to look at solid xenon, which melts at $-169°$ Fahrenheit. Stronger solids are held together by van der Waals forces between molecules, but they get help by having the molecules themselves held together by covalent bonds. Ordinary ice is a crystal held together by the weaker ionic bonds between the hydrogens of one molecule and the oxygens of others (which are often called hydrogen bonds).

Molecular mechanics describes what goes on in and between molecules at this level of detail. A particular set of parameters describing the strengths and behaviors of the forces and bonds in molecular mechanics is commonly called a force field or sometimes a potential, short for "potential energy." (The potential energy of two atoms with a force between them varies with their distance, just as the potential energy of a heavy object increases with its height above the ground.) Potential and force are two sides of the same coin; force is just the derivative (steepness of slope) of potential. In studying the same molecule, a chemist is likely to be more interested in the potential and a mechanical engineer more concerned with the force.

Van der Waals and ionic forces are pairwise and radial; that is, they cause forces between two atoms to push them directly toward or away from each other. Covalent bonds are more complex. They have a radial, pairwise component that is typically called "stretching."

Then a component called "bending" depends on the angle between two bonds on a given atom. A component called "torsion" depends on the angle between the planes formed by three bonds (connecting four atoms). The actual magnitude of the force depends on which elements are involved, and sometimes on what other elements they are bonded to. For example, the length of a carbon-carbon bond depends on whether the carbon is bonded to three other carbons, as in graphite, or four carbons, as in diamond. Various formulations of force fields have other bells and whistles, but these are the main features.

Most force fields were developed by chemists for different reasons, using different methods to arrive at their parameters. Many are fine-tuned to work best in the aqueous environment of a cell's interior and are used for analyzing DNA, proteins, drugs, and so on. Others are optimized for the study of crystals, metals, plastics, and so forth. The reason tuning matters is that the force field is, after all, an approximation to the true quantum mechanical situation, and in a different situation a different approximation may be a little closer.

To a chemist, "a little closer" makes a big difference. For example, a very small difference in the potential of a protein folded into shape A as compared with shape B can mean that in the cell or test tube, virtually all the protein molecules will be in shape A and almost none in shape B. Protein molecules are floppy, with many places where just one bond joins one part of the molecule and the next. The amount of energy necessary to cause that bond to turn one way rather than another is very small. The nanomechanical engineer, however, can use the difference between science and engineering to good effect. We can design stiff molecules where there is at least a tripod of bonds holding each atom in place. The amount of energy necessary to change the shape of such a molecule is much greater. This means that if there's a small inaccuracy in the force field, it hardly changes the shape of the molecule at all, where it would make the protein completely different.[3]

For example, the molecular mechanical engineer can use, for preliminary work, a force field that only describes bond stretching and bending. Bond torsion is a correction of about 1%, and thus can be ignored for a first approximation. Any molecule where bond torsion would make a significant difference is nowhere near stiff enough for use as a molecular machine part. In contrast, solution chemists often use an approximation where the bond stretch isn't modeled, but an average length used as if it were completely rigid. They can do this because in solution, there are no forces strong enough to change the average length significantly.

Imagine simulating a steel chain. If you're a chemist, you're trying to predict what shape of a heap it will make when you drop it on the floor. You need to be very exact about how much friction there is between links, how they bounce off one another as they clatter down, and so forth. You can assume each steel link will remain the same size and shape, though, and worry only about the smaller forces that move them around. If you're a molecular engineer, however, you're simulating the chain being used to lift a ten-ton weight. You can bet that the links will hang in a straight line from the winch to the weight; but you had better be concerned about how they stretch and deform as the weight nears their holding capacity.

The molecular mechanics force field, together with the description of the particular molecule you're interested in, lets you calculate a lot of things about it. It will tell you what shape it will take, and how hard it will be to bend, and what its vibrational frequencies will be. The other thing you can do is simulate it in what is called molecular dynamics.[4]

In molecular dynamics, you take a molecule or a set of them, and give each atom a velocity, chosen at random, to simulate the kind of velocities they'd have due to heat at, say, room temperature. Then you use the force field to calculate the force on each one, and figure where it would have moved to a femtosecond or some fraction thereof later. Now make a movie, doing this entire calculation once per frame.

Watching this movie lets you see if the machine you designed does what you thought it would. Molecular dynamics is peculiarly suited to simulating nanomachines as contrasted with many of the other things, like protein activity, that it is used for. The reason is that things happen much faster in a nanomachine, and the glimpse you get with a simulation is brief.

A femtosecond is an unimaginably short amount of time. If you made a movie in which each successive frame represented a femtosecond of simulated time, it would take a million years to watch one second of simulated time. The speed of sound in diamond is faster than ten miles per second, though, so it would only take two seconds of viewing time for a vibration to move across a one-nanometer part.

Most existing molecular dynamics software, and the force fields it implements, is aimed at chemists and materials scientists. That is, it typ-

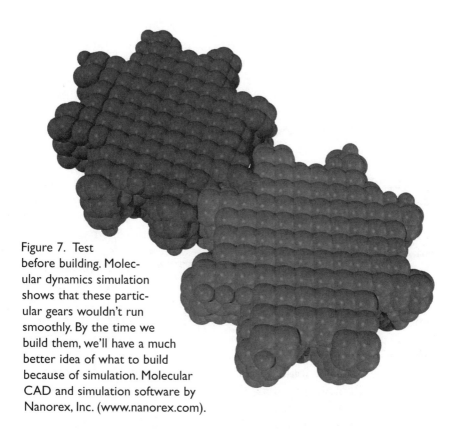

Figure 7. Test before building. Molecular dynamics simulation shows that these particular gears wouldn't run smoothly. By the time we build them, we'll have a much better idea of what to build because of simulation. Molecular CAD and simulation software by Nanorex, Inc. (www.nanorex.com).

ically treats the group of atoms it's simulating as a sample of a much larger amount of material, and it is used to determine things like heat capacity, pressure-volume-temperature relationships, and so forth. For simulating nanomachines, however, we want to treat molecules as machine parts and measure such things as stiffness, torque, and power. New molecular mechanics and dynamics software is being developed specifically for the purpose of simulating and analyzing mechanical nanomachines.[5]

NANODESIGN

Nanomachines are going to be among the most complex things that humans have ever designed. Grains of beach sand range in size from about 100 microns to about a millimeter. A nanomachine the size of a small grain of sand might easily contain five trillion parts, many of which would be moving parts. If you spread the parts out on a table to get a look at them before trying to put the machine together, and you magnified them only to the size of grains of sand, your table would have to be bigger than 500 square feet. If you magnified them to the size of typical machine parts in a car, the table would have to be bigger than 100 acres.

VLSI (very large scale integration) chips such as microprocessors are among today's most complex designed machines, with millions of working parts. Even these are much too complex to design by hand, even for a large, well-organized engineering team. Instead the team relies heavily on design and simulation software. There are a number of strong parallels between microchips and nanomachines. First, you can simulate them at many different levels of detail and accuracy, with more accuracy being more computationally expensive. On a chip, you can simulate the transistors and wires as three-dimensional pieces of material and simulate the electron density flowing through them. Once you have transistors that work the way you want, you tend to have simpler models for them and the wires. You can simulate the generation of electromagnetic waves by the flowing current or ignore it. You can treat the transistors as simple on/off switches, and either simulate or not the delay times for signals to move through them. You can simulate just the logic, where small groups of transistors are represented as Boolean functions such as AND and OR, the primitive one-bit operations of which the computer's logic is built. Even at this ultrasimplified level, a microprocessor is fiendishly complicated.

Each of these levels of simulation will properly represent the machine if the science/engineering distinction is observed. In other words, each level of simulation makes some assumptions about the design at the levels below it and only gives valid results if those assumptions are true. It's up to the designer to make sure that they are.

In electronics design, the assumptions are ensured in a couple of ways. First, designers often make use of predesigned parts that have been carefully analyzed at the lower levels. For example, someone doing logic design will use logic gates that have been carefuly checked out as electronic circuits to have guaranteed limits on signal rise and fall times, propagation delay, and so forth. Then the logic designer need only be concerned with the Boolean logic and not have to worry about electronic engineering as well.

Second, the CAD software that designers use contains subsystems called design-rules checkers. These are very similar to the spelling checkers that are part of text-editing programs. If you're drawing a circuit as a picture on your screen, a design rule may say you can't put two wires closer than some given distance apart. You don't need to know the low-level analysis that showed inductive coupling if they were too close, or that the diffraction limit of the fabrication process just won't let you actually build them closer. All you need to know is that when you draw them too close, they get outlined in red boxes and you must move them apart.

With libraries of prepackaged modular parts and automatic design rules checking, you can be confident that circuits designed at quite an abstract level will work as intended. Exactly the same techniques will be used to design nanomachines.

Today, when people design physical mechanical machines, they pay a lot of attention to each part, optimizing function, manufacturability, and cost. This is very much like how integrated circuits were designed in the 1960s and 1970s. There was enormous attention to detail, and lots of analysis of the effects of each part on every other part. Then circuits began getting more and more complex, as integrated circuits (ICs) became large scale integration (LSI), which became very large scale integration (VLSI). The complexity of a chip rapidly hit and surpassed the limits of what a human brain could understand. A semiserious joke in the industry in those days was that any electronics engineer had two chips in him—after he'd designed two, he was burned out and fit only for management.

There are three broad ranges in the complexity of machines. The first is like a person living alone, with a bedroom, wardrobe, kitchen, garage, and car. Everything's simple and straightforward: when you need the car, you drive it; when you need to cook, you go to the kitchen. The second stage is more like having a large family living in the same circumstances. Now everything is being used more efficiently, at the cost of keeping track of trade-offs and interactions. Someone's always in the kitchen, and Junior is doing chemistry experiments in the sink between meals. Dad gets driven to work because Mom needs the car during the day. You almost need a reservation to use the bathroom.

The third stage is like a large company. Each person gets facilities that are appropriate to his task. Sharing and interaction are planned and are nowhere as helter-skelter as in the family. There are what seem to be large inefficiencies: the office building sits unused all night, the cafeteria is unused much of the day, where with the family, homework is done in bedrooms and on the kitchen table. Still, everything works well enough.

In electronics, the first stage is the discrete transistor in its own package. In mechanical engineering, it is the craftsman building a single tool. The second stage is the IC of the 1970s or the mechanical system of today. Lots of ingenuity goes into getting everything into the constrained resources and making it work together. The giant VLSI chips of today, and the astronomically complex nanomechanical systems of tomorrow, represent the third stage. In the third stage, the bottleneck is not the physical resources of the system, but the ability to design and direct. In the company, it's often cheaper to get extra rooms, machines, whatever, than to buy fewer ones but hire another manager to decide how they are to be shared efficiently. In VLSI, modules are predesigned and used as is, even though each one might have been further optimized for its specific function. Chip layouts are regularized and laid down by automatic programs, even though careful (and time-consuming) human attention might have done it a little better. Indeed, chip designs today look more like programs, written in languages like VHDL and Verilog, than they do any graphical picture of the chip.

Stage three for nanomachines will follow the same route. A mechanical engineer of today, looking at (any small part of) an advanced nanomachine, would consider its design quite sketchy and pedestrian. The reason is that it will be generated automatically as part of a vast system, probably described in something like a programming language instead of anything three-dimensional. The designer will have written, essentially, what she wanted the machine to do, rather than trying to describe the low-level mechanism to do it.

Indeed, the designer might not even know (or care) whether she is describing hardware or writing software. Suppose she is designing a machine that will perform a certain motion. She describes the motion she wants, and the design software decides whether it will be a clever linkage, a more straightforward linkage driven by cams, a general arm driven by stepper motors controlled by a microprocessor, or some combination.

Several kinds of current-day software have capabilities that promise the kind of design capabilities that we will need for big nanomechanical systems. There are the VLSI design systems.

Another, more common example is software development systems. Back in the 1970s, Fortran and Cobol were the tools many developers used in scientific and business programming, respectively, but a preponderance of work was still done in assembly language, specifying low-level machine instructions one at a time. Today, a vast array of software synthesis techniques is available. Specialized programming languages for almost any imaginable purpose are available, which cut the effort to develop a program by orders of magnitude compared to assembly language. For highly stylized applications, systems that are essentially automated questionnaires let you describe the application you need, and generate it automatically. And there are methods at virtually every point between these extremes.

SIMULATION AND TESTING

Today, the best we can do to test a nanomechanical design is simulate it. In the future, however, it will be possible to build and test systems of considerable complexity, like the trillion-parts grain of sand, in seconds. Design practice has tended to move from physical testing to simulation over the past several decades, as computers have become cheaper and faster. In the next few decades, however, the ability to make things physically will follow the same track. For certain applications anyway, there might be a swing back in the opposite direction.

Imagine a complete machine shop, factory, and engineering laboratory in a building with ten acres of floorspace. It has every conceivable kind of construction and test equipment, overhead cranes, mobile robots—the works. Reduced to the scale of nanomachines, this fits in a 40-micron square. It would sit comfortably on the cut end of the finest human hair. Your nanoengineering design workstation could have thousands of these facilities built right into the processor. Having specified your design, the workstation could build and test thousands of minor variants in the time it took to press the return key.

CHAPTER 5

NUTS AND BOLTS
The Basics of What a
Mature Nanotechnology Will Look Like

Inventor, n. A person who makes an ingenious arrangement of wheels, levers and springs, and believes it civilization.
—Ambrose Bierce, *The Devil's Dictionary*

N anotechnology is not a set of particular techniques, devices, or products. It is, rather, the set of capabilities that we will have when our technology gets near the limits set by atomic physics. We can make predictions for such a technology without knowing the specifics of how it will be achieved. We can, for example, know the strength of a substance with a given pattern of atoms and covalent bonds without knowing the process by which it was formed.

We do have to know, however, the pattern of atoms and bonds. The laws of physics don't tell us directly how strong a material can be; they tell us how strong a particular one will be. It's similar with other things of interest to a technologist. Physical law doesn't tell us how powerful a motor can be; it lets us say how powerful a specific motor will be. So we can only get a grasp of the outlines of the capabilities of nanotechnology by analyzing a set of designs. So we have to come up with a set of designs.

These designs have to be simple enough so that we can analyze them now. The more they look and act like machines we already know about, the less predicting we have to do and the more likely we are to get it right. Note that this is not necessarily the way many actual nanomachines will be built, and almost certainly not the most efficient, powerful, or economical way to build them. But if we can design nanomachines like current-day macroscopic ones, we know all the things that need to be analyzed, from structural strength to frictional heating, and we'll have good reason to believe that they will work. That means that the real nanomachines of future designs will almost certainly be better than our designs here. We know for sure that they'll be at least as good.

Another property that we would like to have in our future technology is autogeny. An autogenous technology is one whose manufacturing base—all the machines that make machines—is capable of producing any piece of machinery in the manufacturing base. In our context, it means designs that would, if they existed, be able to build more more machines like themselves. An example is today's machine tool industry: it consists of machines of steel cut into shapes with a ten-thousandth of an inch accuracy, and it can make machines of steel cut into shapes with a ten-thousandth of an inch accuracy. It's a stable point in the panorama of machine-building possibilities.

SCALING LAWS

An ant doesn't look like an elephant. An ant can lift hundreds of times its own weight, but an ant the size of an elephant couldn't even stand up.

The reason is scaling laws. The one directly applicable to the ant/elephant question is called the square/cube law. The weight of an object varies with its volume, that is, with the cube of its length, where the strength of its legs varies with its cross-sectional area, and thus the square of the length. In other words, an ant the size of an elephant, one thousand times as long, would have legs a million times stronger, but it would weigh a billion times as much.

If we run in the opposite direction, down to nanomachines, it turns out we can just about ignore weight entirely.[1]

Don't try to bend a windowpane of glass in your hands. It won't bend enough for you to notice, and if it breaks you'll cut yourself. But an optic fiber, and the fibers in the fiberglass fabric of an auto/boat patch kit, bend easily. You could tie knots too small to see in the ultra-fine fibers of fiberglass insulation. They're all the same stuff: glass. Being thinner means being more flexible—another scaling law. Parts in nanomachines are almost inconceivably thin, and so we'll have to use the stiffest materials we can, like diamond, to compensate.

Swing your arm in a full circle, taking one second to do so. Your hand will move at between one and two meters per second. A robot arm a million times shorter than yours would make a circle a million times shorter, so if it moved at the same speed, it'd go around a million times per second.

Of course, if you tried to swing your arm around a million times per second, it would fly right off. Your bones and tendons are way too weak to support that kind of force. But the scaling laws say that the tiny robot arm will experience the same stress per unit area (in general) as yours when its hand moves at the same speed. So its frequency can go up as its size goes down. You can see the same phenomenon at work in the animal kingdom. Pelicans and geese flap their wings no faster than you can wave your arms. Sparrows and wrens flap much faster, and a hummingbird's wings are a blur. (If you can see even a blur around a fly in flight, you have sharper eyes than I.)

You could pick up and place, with reasonable accuracy, small items from one spot on your desk to another at about one per second, but a nanoscale robot arm should be able to do the same at about a million per second. While car engines run at thousands of revolutions per minute (RPM) and small electric motors at tens of thousands, nano-engines and motors can turn at billions or tens of billions of RPM, well within conservative design parameters.

STRUCTURES

Today's macroscopic machinery is mostly made of metal, although plastic of various kinds has played an increasing role for the past half-century. Steel is king, with aluminum alloys playing the role of prince. But we can't make nanomachines from steel, or aluminum or plastic either. It's those scaling laws.[2]

The same mathematics of form that make fiberglass flexible means that a beam of steel at the nanoscale would be about as useful as one of Play-Doh in the everyday world. To make useful beams, frameworks, casings, and shafts, we need to use the stiffest, hardest stuff we have: diamond. (Some theorists have claimed that a structure of carbon nitride is harder, but no one's managed to make it yet.) Even diamond is a bit rubbery at the nanoscale. It would be nice to have something harder, but diamond will do.

As I write, it is diamond's golden anniversary: synthetic diamond, that is. Just fifty years ago, diamond, real honest-to-goodness diamond, of the same molecular structure as natural gems, was first made in the laboratory.[3] Today, it's made by the ton in factories: about a billion dollars' worth of synthetic industrial diamond is sold each year.

Nature produces diamond by compressing carbon at extreme temperatures and pressures. The first industrial method for diamond synthesis worked much the same way, producing spoonfuls of diamond grit in gargantuan one-hundred-ton hydraulic presses. But the way most diamond is made today is by chemical vapor deposition, or CVD. This is a method whereby, at relatively low temperatures and pressures, one can grow diamond crystals from carbon-bearing gas such as methane (natural gas) or acetylene. Even alcohol has been used.

Some reactions are very finicky. CVD isn't. Robert deVries, retired General Electric diamond researcher, quipped, "You can almost make this at home with a microwave oven."[4] The process has been duplicated by amateurs in their garages. In any chemical process, you can control the conditions only on the average. The actual reactions involve random collisions between the gas molecules and the diamond being formed. But in a nanomachine, everything is precisely controlled. You can specify just which molecule touches which, when, and how hard. Robert Freitas and Ralph Merkle have made an extensive study of the process.[5] Given how easy diamond making is with random gas collisions, it's unimaginable that it won't be quite straightforward in a nanomachine.

Diamond is a wonderful building material. It's five times as hard and twenty times as strong as steel. (That's high-alloy steel. It's one hundred times as strong as the "mild steel" that coat hangers are made of.) It's as slippery as Teflon. It's a good conductor of heat and a good insulator of electricity, both excellent qualities in a framework material. And its thermal expansion is tiny. So we'll be using a lot of diamond (and other gemstone or gemlike materials: sapphire, tungsten carbide, boron nitride, etc.) in our nanomachines. Cylindrical chunks of diamond work fine for shafts, and slabs of it work fine for walls and supports.

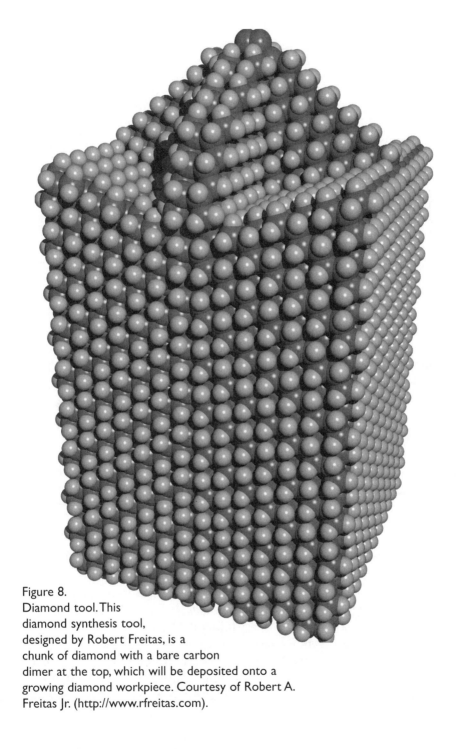

Figure 8.
Diamond tool. This
diamond synthesis tool,
designed by Robert Freitas, is a
chunk of diamond with a bare carbon
dimer at the top, which will be deposited onto a
growing diamond workpiece. Courtesy of Robert A.
Freitas Jr. (http://www.rfreitas.com).

BEARINGS

The most important element of today's machine technology is the bearing. The simplest thing a part can do repeatedly is turn, and bearings allow it to do this while supporting a force. In your car, the crankshaft, driveshaft, axles, and wheels all turn, supported by bearings. If the bearings fail, the power of the engine is converted by friction into heat instead of motion, and the car seizes up or catches fire.

The simplest form of a bearing is a sleeve bearing, simply a shaft going through a hole. The harder the material of the shaft and hole, the better; that's why jeweled bearings are used in fine watches. A door hinge is a common example of a sleeve bearing. Like any macroscopic bearing, a hinge turns more easily with a few drops of oil on it.

At the nanoscale, a molecule of oil is a lot more like a loose part than a lubricant. On the other hand, we can do something unavailable at the macroscale, which is to make the bearing surfaces atomically precise. That's as smooth as the laws of physics allow. They're still "bumpy" with individual atoms, but atoms are soft and slippery. A common lubricant at the macroscale is graphite. Graphite is a form of carbon that comes in stacks of sheets, where each sheet is a single molecule, atomically smooth. The sheets slide over each other easily. At the nanoscale, we can roll the sheets up into seamless tubes. Two such tubes, nested inside each other, should make a very nice bearing.

They do. Such tubes, called buckytubes or fullerene nanotubes, can be made now, on a hit-or-miss basis, in enough quantity to be experimented on in the laboratory. Two of them nested one inside the other are sometimes called a double-walled nanotube (DWNT). Such nanotube bearings have been demonstrated in the lab, with the outer tube rotating on the inner and carrying a load.[6]

I'm sure that lots of graphite bearings will be used in nanomachines. However, they have a couple of drawbacks. First, they have no resistance to an axial force: the tubes are just as happy to slide in a linear fashion as to rotate. For some applications, this is just what you want. Consider the brake cables on a bicycle. For others, it isn't. You don't want the axles on your car sliding back and forth.

The other problem is that graphite is soft. In many applications in a machine, you want the bearings, and everything else, to be as hard and stiff as possible. If your door hinges were made of rubber, the door would sag, scrape the floor, and hit the jamb. The stiffer the

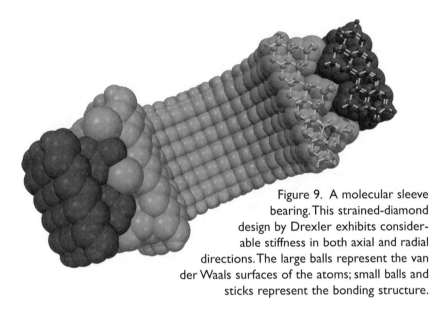

Figure 9. A molecular sleeve bearing. This strained-diamond design by Drexler exhibits considerable stiffness in both axial and radial directions. The large balls represent the van der Waals surfaces of the atoms; small balls and sticks represent the bonding structure.

bearings (and the parts), the more likely the parts will go where they were planned to.

We can design (but not yet build at the nanoscale) bearings of diamond and similar hard crystals, though jeweled bearings have been used in fine watches for centuries. With careful attention to the configuration of surface atoms, these can be quite slippery to rotation, but resist axial and radial forces better than graphite ones.[7]

Macroscopic sleeve bearings are cheap and ubiquitous, but for more demanding applications, more sophisticated forms of bearings are used. One major reason is friction. There is no friction at the atomic scale, but there are effects that cause some of the same problems.

At the macroscale, friction causes wear. A properly designed nanomachine will not wear. Wear occurs because the surface imperfections of macroscopic parts, or grit, dislodge material when surfaces move in contact. A nanomachine, if properly designed and built, has no surface imperfections and no grit. (This doesn't mean that a nanomachine will last forever. One good hit from a cosmic ray and it's a pile of junk.)[8]

Heating, the other effect of friction, is something we do have to worry about. Atomically smooth doesn't mean mathematically smooth. So when two surfaces made of atoms move past each other, the atoms of one will press alternately against the atoms of the other and the gaps in between. Now, there is no dragging force as one atom slides on another; atoms themselves are perfectly smooth. But an atom moving across a bumpy surface gets bounced up and down, and that bouncing is heat. Rub your fists together, with your knuckles in contact. You'll feel the vibration all the way up your arms. That vibration represents lost energy in a nanomachine.

You can cut down this vibration a lot by giving the two surfaces in contact a different pitch, so that one has a different number of atoms per nanometer than the other. You can get an intuitive feeling for the difference by sliding a couple of machine screws along each other. Two screws of the same pitch will lock up as the grooves match; screws of different pitch will slide. Even so, there's still some energy dissipation from sliding surfaces.

At the macroscale, more sophisticated bearings use balls or rollers. Surfaces don't slide, they roll. The same techniques work fine at the nanoscale. You have to remember that even diamond is rubbery at the nanoscale; rolling across it is like rolling across a box spring. You can lose energy by making the whole thing jiggle. But that can be minimized as well, especially if the bearing is designed to run at one specific speed. The natural vibration of the part can be tuned to feed the energy back into the mechanical motion rather than dissipating it as heat.

The next machine element we need is the gear. Here the bumpiness of atoms is an advantage. It's quite possible to have gears whose teeth are atoms or rows of them. In this application, as with macroscopic gears, the important thing is to get the pitch to match so they *don't* slide easily.

Since atoms are soft and diamond is rubbery, you can—in fact, you must—press gears together, where in a macroscopic machine this would cause jamming. In a nanomachine, however, this means that the gears are supported and fewer bearings are necessary. In fact, for some designs, there's no distinction. A planetary gear and a roller bearing can be the same thing.

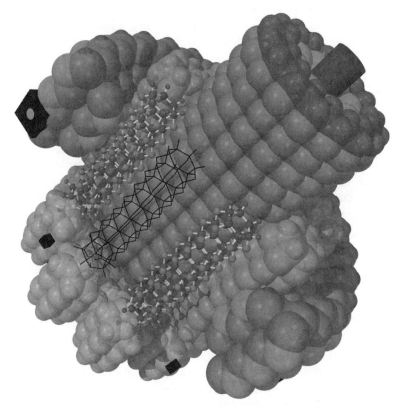

Figure 10. A planetary gear. This early version of a planetary gear by Drexler was the first nanomachine system to be simulated with molecular dynamics.

MOTORS

The subject of fueled engines deserves its own chapter, but we'll consider electric motors here. Electric motors at the macroscale work by magnetism. Current flowing through wires creates magnetic fields that exert physical forces between the wires and other current-carrying wires or permanent magnets.

Unfortunately this won't work at the nanoscale. It's those scaling laws again. The magnetic effect gets less powerful as the scale goes down. Luckily, the electric effect gets more powerful, so we'll use it instead. The electric effect is powerful enough to cause motion even at the macroscale—it's what causes static cling in clothing. At the nanoscale it's almighty powerful. In fact, it's what holds atoms and molecules themselves together.

The simplest way to design an electrostatic motor is much like a waterwheel: electrons are pumped in at the top, they fill buckets on the wheel, and they're sucked off at the bottom. Instead of gravity pulling them down, the electric force pushes them from a low voltage area to a high voltage one (low to high because electrons are negatively charged—blame Benjamin Franklin).

The power of a motor is its torque (rotary force) times its speed. For this kind of electric motor, ignoring friction and the like, power is also the voltage times the current. In fact, the torque is proportional to the voltage and the speed determines the current: the faster those little buckets go, the more electrons are being carried across per second.

Drexler, in *Nanosystems*, analyzes a motor of this type.[9] Its wheel is 390 nanometers in diameter and 25 nanometers thick. It rotates at 800 megahertz (48 billion RPM). The buckets of electrons are moving at about Mach 3 (faster than 2000 mph—the interior of the motor must maintain a vacuum). It draws 110 nanoamps of current at 10 volts, and produces 1.1 microwatts of power.

This may not seem like much, because the motor is tiny. A billion of them, producing 1.1 kilowatts, would be about the size of the ball at the point of a ballpoint pen. Enough of them to match the 100,000 horsepower of a large commercial jet engine would fit in the palm of your hand.

Oh, and by the way, Drexler's design is overly bulky, since he was designing a combined motor and flywheel. The motor could be designed to work fine in at least fifteen times less volume. The bottom line is that for any macroscopic device using nanotechnology, motors will never be seen. Microscopic motors can be strewn throughout the fabric of whatever you're designing, capable of providing more physical force or power than you could ever need.

For applications where a precise position is needed instead of top speed and power, a stepping motor, similar to a clock motor, is indicated. Like other motors, macroscopic steppers are electromagnetic but the nano-sized ones would be electrostatic. One way to build such a motor would be to separate electric charges in the rotor (this is easy—even so common a molecule as water has a built-in charge separation) so that its rim has alternating areas of positive and negative charge. Then surround it with electrodes that you charge and discharge with alternating currents. Driving this kind of motor is a lot more complex than the other kind, but you can make it go forward, backward, stop, turn a specific angle, and so forth. Stepping motors are commonly used today in robotics.

FROM HERE TO AUTOGENY

Now that we have motors, shafts, gears, and bearings, we can build a robot arm. An arm is basically just a series of boxes on hinges with a motor to power each joint. Typically the motors are geared down to provide more force and higher precision (and stiffness) than the motor would have alone. Robot arms with grippers (a finger is just a miniature arm) can pick up things and move them, and in particular can build more robot arms.

But where will it get the parts?

Macroscopic parts are made in many different ways, ranging from molding, to forging (beat it into shape), to machining. Machining is the most expensive, often the slowest method, but also the most accurate and least restricted in what it can make. A milling machine and a lathe can make just about any machine part (you need special attachments for screws and gears).

Milling machines and lathes can be thought of as very specialized robot arms. What they do is hold a knife and carve the shape you want out of the chunk of material you give them. To get the shape just right, they must hold the knife very firmly. This is technically known as stiffness, and it tells you how far the knife (or whatever) will move given a certain force against it. The higher the stiffness, the better.

A milling machine doesn't look like an arm. All the joints are big, heavy, precision-machined hunks of metal that don't wobble at all. If we build a nanomachine on the plan of a milling machine, the position of the tool could be very exactly controlled with respect to the position of the workpiece. And this is the kind of machine that we'll make to build the parts that the slender robot arms will put together.

There's just one big difference. A classic milling machine (or lathe) operates by taking a chunk of material and cutting off everything that doesn't look like the part you need. At the nanoscale, a better strategy would be the opposite: to build up the part you want, an atom or molecule at a time. This is what allows you to have complete control over where every atom is in the finished product.

This method of controlling the reaction precisely is called positional chemistry or mechanosynthesis.[10] Mechanosynthesis is what will allow us to make million-atom parts and get every atom exactly right. This, in turn, will allow us to build molecular milling machines and do mechanosynthesis. And that's our autogenous technology.

PUMPS

Unless you live on an old-fashioned farm and milk your cows by hand, the milk you drink came through a milking machine and subsequent processing and handling stages, courtesy of a number of pumps. Unless you own an artesian well, the water you drink was pumped, either from the ground or by way of water mains from a reservoir. Commonly used fuels are pumped, many times, through processing plants, from one tank to another, along pipelines, into your car at the service station, and again from your gas tank to your car's engine.

Pumps for air in low-pressure applications are called fans. There are probably fans in your computer to cool it, fans in your home heating system to heat and cool you, and more fans in your car. A typical hydronic home heating system has both fans and pumps.

Moving fluids around and managing their pressure is a substantial application of, and a necessity for, modern technology. Nanomachines will need to do it, too.

If anything, most of the designs for pumps at the macroscale will work better at the nanoscale. The simplest pump is a fan. Fans are noisy and inefficient because they cause turbulence. At the nanoscale they won't, because the smaller you get, the more syrupy fluids act. For gases, fans can be made that interact more or less individually with the molecules. Molecules of a gas bounce off one another like billiard balls, and fly some distance before they hit another molecule. The average distance for a given gas at a given pressure is called the mean free path. If the length of a molecule's path through the fan is smaller than the mean free path, molecules going through the fan tend to interact with the fan and not one another. If the fan blades can move faster than the molecules tend to, the fan can attain something

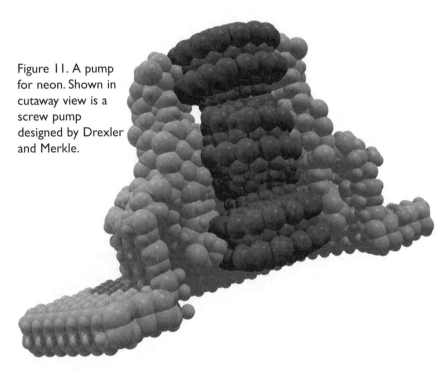

Figure 11. A pump for neon. Shown in cutaway view is a screw pump designed by Drexler and Merkle.

like a compression ratio of ten with one rotor and one stator (or pair of counterrotating fans). Ordinary air has a mean free path of about 60 nanometers and an average speed in the neighborhood of the speed of sound—figures easy for a nanofan to match.[11]

One common pump for liquids, the centrifugal pump, is not recommended at the nanoscale for scaling-law reasons. But the kind you probably think of when someone says *pump*, the piston or positive displacement pump, should work fine.[12] The positive displacement pump usually consists of two one-way valves and a cylinder in which a piston goes up and down. At the macroscale, the valves are often spring valves, but at the nanoscale it will probably be preferable to make them mechanically actuated in a timed sequence with the piston, like the valves in a car engine.

Can we make a positive displacement pump that interacts with the fluid one molecule at a time? Yes, and in a fairly elegant design. Consider a cylinder with a set of molecule-shaped dents in it. It is embedded in a wall like a revolving door. The molecules go through as the cylinder turns, just as people go through the door. But the trick

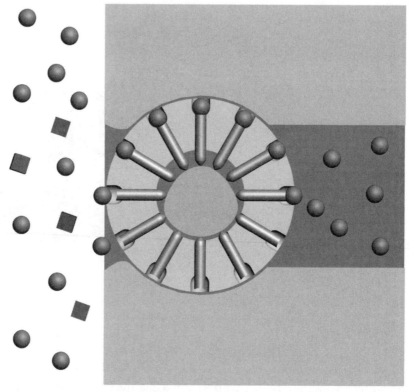

Figure 12. A sorting rotor. This schematic view shows a sorting rotor pumping one fluid out of a mixture. Like any pump, it would require power to operate. The cubes represent molecules that wouldn't fit into the pockets in the rotor. Courtesy of Gina Miller (www.nanoindustries.com).

is that a little piston at the bottom of each dent pushes the molecule out when it's on the output side, and retracts to let a new molecule in on the input side.

Doing it this way allows you to make the dents much more attractive to one kind of molecule than to another, by making them fit snugly. This is how a large proportion of the molecular machinery in your body works—it's how hemoglobin grabs oxygen molecules to carry them in your red blood cells, for example. With a set of dents that strongly favors one kind of molecule, we can pump that fluid preferentially out of a solution—for example, oxygen out of air, fresh water out of seawater. This kind of selective pump is called a molecular sorting rotor.[13]

NANOCIRCUITRY

Light and radio are waves in the same stuff: electromagnetic fields. The only difference is the length of the waves, and thus the frequency. With fast enough nanoswitches and appropriately sized antennas, we should be able to handle light with the same facility that we now handle radio waves.

Today, for example, we have phased-array antennas for radar. A flat array of antennas can produce a beam going in any desired direction by having each little antenna match the phase properly. What's more, it can detect the direction of an incoming beam without a lens or mirror. In essence, it's a controllable hologram at microwave frequencies.

With nanoelectronics and nanoantennas, we should be able to do the same thing with light—phased-array optics.[14] Cover the surface of an object with optical antennas and it should be able to take on any properties you like that involve the reflection and absorption of light, or generate a completely synthetic pattern, making it appear to be anything you want. You could even have it present the image that would be there even in its absence, making it effectively invisible.

In order to make antennas and to connect switches, you need wires. Macroscopic wires are generally made of metal. Metal conducts electricity because the electrons (some of them anyway) aren't nailed down to particular nuclei or in particular bonds, as they are in diamond, for example, but are free to roam around at the behest of any passing electric field. Graphite conducts electricity because electrons can roam, but only within a given sheet. Rolled-up graphite, or buckytubes, can conduct or not depending on their particular structure. Various molecules have areas in which electrons can move around. These areas act as optical antennas at various frequencies, which is why different substances have different colors and albedos.

Conductive paths at the nanoscale can be made in many ways besides laying down wires of metal, but wires of metal do work. The other ways just give us other options.

Some arrangements of atoms form superconductors—substances

with no electrical resistance at all. Since superconductors can be made with the statistical mixing of current-day chemistry, it seems likely that we could make better ones with a technology that allowed us to place atoms in a specific, desired pattern. Superconducting wires would be useful at the nanoscale, since ordinary wires gain more resistance as they get thinner. It remains to be seen whether superconducting wires can be made thin enough to be useful, however.

PUTTING IT ALL TOGETHER

Most of the elements—parts and mechanisms—of macroscopic machines can be scaled to work at molecular size, like sleeve bearings, or redesigned to, like electric motors. The few that can't, like centrifugal pumps, have substitutes, plus there are many new designs, like sorting rotors, that don't have any macroscopic parallel at all. We can confidently expect to be able to design and build machine systems at the nanoscale with a wide variety of capabilities and applications, and in particular, to design and build manufacturing systems.

CHAPTER 6

ENGINES
Making Molecular Machines Go

Gottes Muhlen mahlen langsam mahlen aber trefflich klein
Ob auss Langmuth er sich seumet bringt mit Scharff er alles ein.

(The mills of God grind slowly, yet they grind exceeding small;
With patience He stands waiting, with exactness grinds He all.)
—Friedrich, Freiherr von Logau,
Deutscher Sinngedichte Drey Tausend

In 1712 the first steam engine was built in Birmingham, England, by Thomas Newcomen.[1] It was an impressive affair, standing three stories high and doing the work of a team of horses. The Newcomen engine had been made possible by the discovery of the atmosphere in the previous century by Galileo, Torricelli, Pascal, and others. It seems odd to us that the atmosphere should need to be discovered; we take it for granted. But the notion that we stand at the bottom of an ocean of air, and that the pressure it exerts on a square foot of surface is the same as the weight of a one-square-foot column of air extending to the top of the atmosphere, would have astounded Aristotle or St. Augustine.

Newcomen's engine worked directly by the action of that atmospheric pressure. The air was pushed out of a cylinder by steam. Then the steam was reverted to water by a cooling spray, leaving vacuum in the cylinder, which drew down a piston.

A horse, by a suitable arrangement of pulleys, can raise a thousand-pound weight thirty-three feet in a minute, at a pace of work that it can keep up indefinitely. Thirty horses can do in one minute the same work one can do in half an hour; in each case the quantity of work can be called a million foot-pounds. Newcomen's steam engine could do 5 million foot-pounds of work for each bushel of coal that was shoveled into it.

Newcomen's engine was a brilliant, world-changing breakthrough in technology, but it wasn't terribly efficient. Essentially by tinkering and optimizing the basic design, by the middle of the eighteenth century, Newcomen engines had been improved to produce 9 million foot-pounds of work for each bushel of coal.[2]

Then came the Scot, James Watt. In January 1769 he patented an improved steam engine that avoided one of the big inefficiencies of Newcomen's.[3] In Newcomen's engine, the working cylinder had to be cooled down to collapse the steam and produce the vacuum, and then heated up again by the new steam at each stroke. Watt introduced a separate condenser that stayed cold, and let the cylinder stay hot. Watt's engine was a major improvement, produceing 28 million foot-pounds of work for each bushel of coal. The steam engine now began to be competitive economically with the waterwheel. While you didn't have to buy coal to run your waterwheel, you did have to own enough length of a fairly decent-sized stream to have a ten-foot drop, and a twenty-foot drop was even better. With the steam engine you could set up shop anywhere.

One of the things that made a big difference in how much work you got out of your engine for a given amount of coal was the precision with which it was made. Better precision meant less friction, less steam escaping uselessly from from joints and valves, and most of all, better reliability. From an almost entirely handmade technology in 1750, the precision and fit of metalwork improved drastically until the early 1800s, when the beginnings of a machine tool industry could be seen.

Also in the early 1800s, a new surge in steam technology happened because Watt's patents expired. Watt simply wasn't interested in high-pressure steam. This is probably with good reason, since the technology in which his engineering instincts were formed wasn't strong or precise enough to handle high pressures at high temperatures. However, by the early 1800s there had been enough advancement of fabrication and metallurgy that a new breed of engineers, notably Arthur Woolf, could sucessfully build engines that pushed the piston with steam pressure rather than sucking it with vacuum.[4] This had the immediate consequence of making the engines smaller; higher pressure on a smaller piston produces the same force. What's more, they could throw away the condenser, allowing used steam to escape to the atmosphere. Woolf's double-expansion pressure engine of 1804 got 56 million foot-pounds of work out of each bushel of coal.

> Shortly after our arrival we went to the coal wharf to see the arrival of the coal wagons which are set in motion by steam machines and which bring coals from the mines at a distance of six English miles from the wharf. It is a curious spectacle to see a number of columns of smoke winding their way through the countryside. As they approach we see them more and more distinctly until at length with a column of smoke we also perceive the wagon from which it ascends, dragging a long train of similar wagons behind it, which gives it the appearance of a monstrous serpent.
> —German traveler to Leeds in 1816,
> in *Turning Points in Western Technology*

The great practical implication of smaller, more efficient engines was that you could put them on vehicles such as boats or wagons. Steam engines brought power not only for factories but transportation. The burgeoning machine tool industry provided precision parts for other machines as well, notably the spinning jenny and the power loom. In less than a century, England went from an essentially agrarian society where everything was handmade, to one with a significant manufacturing sector, using mechanical power for mass production. Railways made freight cheap and fast enough to produce an integrated national economy. This was the industrial revolution.

Meanwhile, the engineers didn't stop improving and optimizing the steam engine. By 1835 in Cornwall, the leading area of steam engineering of the day, engines could produce 125 million foot-pounds of work for each bushel of coal.

NO FREE LUNCH

Given this history in the improvement of the efficiency of steam engines, it was natural for the more philosophically inclined of engineers to wonder whether there was a limit to the process. A young

French engineer named Sadi Carnot was among them, and in 1824 he published a short book entitled *Reflections on the Motive Power of Fire.*[5] With this book, he founded the science of thermodynamics.

Carnot systematized the knowledge about heat engines first by pointing out that they work from a temperature difference, not just a source of heat: an engine sitting in an ocean of steam couldn't draw any power from it. This had an important implication: any heat flowing across a temperature difference in an engine was a source of inefficiency. If you used a 1,000-degree fire to heat 500-degree steam, you could, in theory, put a whole separate engine between the fire and the steam, getting extra energy out of it, and still run your original engine. So the efficient engine must transfer heat without a temperature drop. This sounds like a tall order, since we expect heat flows to be associated with temperature differences. But it can be approximated by having the working fluid and the heat source in contact at the same temperature, and expanding or compressing the fluid in a way that would normally heat it or cool it; the heat will flow in or out instead.

For example: Stopper a bicycle pump and push the plunger in. It will get hot. Stick it in a bucket of water, and the heat will flow from the hot pump to the cold water. Now take it out and let the plunger up. The pump will get cold. Stick it in the water again, and heat will flow back into the pump. However, the amount of work you got back, when the plunger pushed its way up, was less than what you put in to push it down.

Now put the pump into the water first. Push the plunger down and you warm the water up a little, but the pump and water remain at about the same temperature. Let go: the pump and water will cool down a smidgen, and you'll get back a better fraction of the energy you put in. (In theory you could get back arbitrarily close to all of it, if the pump were frictionless and you moved it slowly enough to prevent buildup of temperature differences.)

Such an isothermal (without temperature drops) heat transfer is called reversible, because, as in our thought experiment, it works both ways without losing energy. The process of compression or expansion can also be reversible while warming up or cooling down, by virtue of insulating your fluid. You obviously don't want your heat leaking off into the environment. The engineer's jargon for a process that is absolutely insulated against heat flow is *adiabatic*. Neither kind of reversible process is possible with any technology like metal cylinders and pistons, but it showed Carnot the way to the theoretically optimal heat engine.

Here's how it works. We have two buckets, one with hot water and one with cold. We want to extract work by moving heat from the hot to the cold. We'll assume our bicycle pump is completely insulated when it isn't in either bucket. Start with the plunger halfway out and the pump in the cold bucket.

1. (Adiabatic compression) Take the pump out of the cold bucket and push in the plunger until it is as hot as the hot bucket.
2. (Isothermal expansion) Put the pump into the hot bucket and let the plunger out halfway.
3. (Adiabatic expansion) Take the pump out and let the plunger the rest of the way out, its temperature falling to that of the cold bucket.
4. (Isothermal compression) Put the pump into the cold bucket and push the plunger in halfway.

Lather. Rinse. Repeat.

You've put work into the pump in steps 1 and 4, but you get more out in steps 2 and 3. The reason: pressure rises and falls greatly during the adiabatic steps, but stays closer to constant during the isothermal ones. Thus step 2 gives you more work out than you put in in step 4, while 1 and 3 more or less even out. Since you ended where you started, you can keep repeating as long the hot bucket is hotter than the cold bucket, getting work out every cycle.

Now, Carnot went on to note, every step is reversible. That means that you could do the whole thing backward, compressing in the hot bucket and expanding in the cold one. You'd put in more work than you got out, but you'd be pumping heat from the cold bucket to the hot one. In other words, a heat engine is a heat pump in reverse; and a reversible heat engine–pump would pump the same amount of heat for a given amount of work, as it would produce work for the given amount of heat.

So, Carnot continued, imagine a heat engine more efficient than a reversible one. That is, one that produces more work for the same amount of heat. Well, just hook it to a reversible engine used as a pump, which can pump back from cold to hot all the heat you used from hot to cold, and you'll get some extra work out besides. In other words, get work out of nothing. And that, as Carnot stated with typical Gallic flair, is "repugnant to science and to common sense."[6]

Thus the industrial revolution dawned. The revolutionary part was the beginning of a process, not the full course of one. Many of the trends set in course by the revolution continue today. One is the improvement of engines. The smaller and more powerful an engine, the more useful it is. The Newcomen engine couldn't be used in its original form for much more than pumping the water out of coal mines. It used too much coal to be economical very far from one. The Watt engines were useful as power sources for factories all over the country; and the Woolf engine was put on a wagon to replace a horse. By 1900, steam engines had become light enough to power a car; the Stanley Steamer gave gasoline engines a run for their money. But for mobile applications, heat engines using air and gaseous combustion products as a working fluid were the clear choice over steam, and have ruled the twentieth century. No useful airplane ever ran on steam. By the middle of the twentieth century, piston engines were being pushed out of the most demanding applications by the gas turbine. Note that steam turbines had succeeded steam pistons in applications like ships through the first half of the century, and are still in use in stationary power plants today.

The reason is Carnot. In a stationary plant, your main concern is efficiency. If weight is no object, you can get just as high efficiency with steam as with any other working fluid. In fact, a steam power plant is notably more efficient than the gasoline engine in your car. However, in a car, light weight is the important thing. It's even more important in an airplane or a spaceship. The apotheosis of engines today is the rocket engine. Each ounce of a modern rocket engine produces more power than Newcomen's entire three-story machine.

Two overlapping sets of trends can be seen in the three-hundred-year history of engine development. First, within the development of any particular kind of engine, a rise of efficiency and power follows an exponentially increasing curve but then levels off as that kind of engine nears the theoretical limits imposed by thermodynamics. The second trend is that a constant rain of new engine types is being invented. For some figures of merit, such as power-to-weight ratio, or size of the smallest, or total power of the largest, and so on, the curves of the different types patch together to produce a single, unbroken, three-hundred-year trend line that is itself exponential.

For example, if we take steam pistons, steam turbines, gasoline pistons, gas turbines, and rocket engines, we can overlay them to get a smoothly increasing power-to-weight curve, stretching from Newcomen to the space shuttle. This is all somewhat qualitative since one must take into account the cost of the engine, how much it pushed the technology, and what other options might have been taken in its stead. But the broad, general trend is unmistakable. The trend says that unless something really drastic happens to make the next thirty years totally different from the last three hundred, we'll have engines with the power-to-weight ratios, sizes, dimensional tolerances, and so forth that nanotechnology predicts.

Horsepower per pound

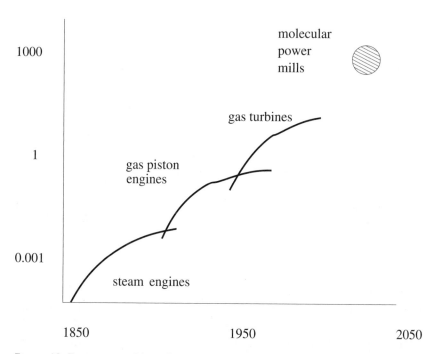

Figure 13. Engines trend line. Going through at least six different technologies—steam suction pistions, steam pressure pistons, steam turbines, internal combustion pistons, gas turbines, and liquid fuel rockets—the power-to-weight ratio of engines has followed a steady exponential trend for nearly two centuries.

Another thing that has been increasing over the years is the efficiency of the engines. This doesn't mean that the laws of thermodynamics are being broken, but that a loophole is being used.[7] The greater the difference between the hot and cold buckets, the more efficient the engine. Advances in materials, and some ingenious tricks like using a curtain of air to keep the flame in a jet engine from actually touching the walls of the combustion chamber, have allowed the temperature of the hot bucket to rise and rise.

With nanotechnology, however, it may be possible to finesse the matter altogether. No law, after all, says that no engine can be completely efficient; some electric motors get remarkably close. They do so by not being heat engines, but by using different principles. They use the laws of electromagnetism to translate voltage differences and current flows into mechanical forces. It turns out that the chemical energy represented by fuel and oxygen is, at the molecular level, just another case of electromagnetic potentials. Thus there's no objection in principle to a device that could convert those potentials to mechanical force.

Such a molecular power mill would, to approach full efficiency, have to be reversible. Carnot says you can't let something hot touch something cold, or an irreversible heat flow will occur. It's the combustion step in current engines that's irreversible, of course; you can't take hot steam, compress it, and get cool oxygen and hydrogen. With a molecular power mill, though, you could put hydrogen and oxygen in the front, the mill would turn a shaft, and room-temperature water would come out the back.[8] The whole process would take place at essentially the same temperature, with all molecules involved in the reaction being held in place. Feed in the water and provide power to the shaft, and you would get oxygen and hydrogen. It is worth noting that many of the chemical reactions performed by the molecular machines of the cell are reversible.

Alternatively, the power input/output of the mill could be electric, like a fuel cell; present-day fuel cells are about 60% efficient, but they are quite heavy. Power mills will represent the far end of the technology curve. Once the reversible mill is achieved, it could be combined with a hydrogen tank for an energy storage mechanism. Current-day car batteries store about one-sixth of a megajoule of energy per kilogram of weight; the best lithium batteries get up to half a megajoule. Gasoline provides 42 megajoules per kilogram, but you get only about 8 out of it with a car engine, the rest being discarded as heat. Hydrogen represents 120 megajoules.

In other words, if you're flying from New York to Miami in a light plane with a gasoline-burning piston engine, you'd need to carry 550 pounds of fuel, but a hydrogen-powered reversible engine would need only 37 pounds of fuel.

There's only one problem: 37 pounds of hydrogen would fill a balloon about the size of Newcomen's original three-story steam engine. It would actually provide 450 pounds of lift! But it would be unmanageably awkward. The hydrogen needs to be kept in a more compact form; so, besides supplying the engine, nanotechnology will be needed for the fuel tank. Currently, hydrogen is stored compressed or liquified. The pressure tanks are heavy compared with the amount of hydrogen they can hold, and liquefaction is out of the comfort range of current technology, tripling the price of liquid hydrogen compared to the gas form. Nanotech cooling and insulating technology could help considerably. (Note that liquid hydrogen is quite light compared to other liquids, so the hydrogen tank would still be the same size as the gasoline tank; but it would weigh only 7% as much.)

It's a lot harder to estimate the size of the engine, but we can get a ballpark figure by looking at estimates for molecular manufacturing systems. A fuel-burning engine is a manufacturing system at the molecular scale: it inputs the raw materials hydrogen and oxygen, and outputs water molecules. Using figures from Drexler's *Nanosystems*, we can estimate that a 100-kilowatt engine, as used by a car or light aircraft, would weigh in at about 50 grams (1.5 ounces).[9]

POWERING THE NANOWORLD

Newcomen's engine, occupying a full three-story building, supplied about as much power as the motor on a present-day lawn mower and consumed so much fuel it could only be used right next to a coal mine. The engines nanotechnology will build will make our current power plants look as bulky and inefficient—and as weak. It's one thing to realize that an engine matching the one in your car could be smaller and lighter than a pocketknife; it's another to realize that your pocketknife, if you wish, can have the power of fifty chainsaws. (And the reach of a chainsaw as well—a foldable, powered, woodcutting blade would be a fairly straightforward design with the materials and machines of molecular engineering.)

An amusing thing happened when I was doing the engineering calculations for the flying cars that appear later in the book. I had made a collection of existing flying machine designs, and a few flying cars from fiction. Generally the serious fictional ones followed the same general form as the real aircraft. A few didn't, seeming to follow the whimsy of the illustrator: the most recognizable of these is the Jet-sons' car from the Hanna-Barbera cartoon. After doing the calcula-tions, I realized that only the *Jetsons* got it right; the engines would actually be the size of the silly little pods shown in the cartoons, if not smaller.

In virtually every application in which motors are used today, those motors will disappear. Tools and machines will still be powered, but the motors will become too small to see, part of the material. Objects will seem to be capable of moving by themselves, like animals. Your toothbrush won't have a motor in the handle, but many motors in each bristle. Instead of a single engine and wheels, a car might have hundreds of small legs, giving you a smooth, rock-steady ride over almost any surface.

Engines will vanish, and will be everywhere.

CHAPTER 7

A DIGITAL TECHNOLOGY
Atoms Are the Bits of the Material World

Charles Babbage is famous today for almost inventing the computer. Specifically, he invented it, and almost got it built, but not quite, and nobody bothered with it again for a century. Once people finally did build working calculating machines, computers took off like crazy. The twentieth century has seen a trend in computer power that is remarkable in the history of technology.

Just like physical power engines, computing engines have stayed on an exponentially improving track through a series of changes in the underlying technology. Computers started with mechanical shafts and gears, as in Babbage's design, followed by electromagnetic relays, vacuum tubes, discrete transistors, integrated circuits, and a series of order-of-magnitude leaps in the size of the circuit that was integrated.

Just as with the steam engines, this leads us to ask whether our computers might be approaching some theoretical limit. The answer turns out to be yes. In fact, it is the same thermodynamic limit Carnot discovered, but in a different disguise.

Modern-day thermodynamics talks about such questions in terms of a property called entropy. Entropy has found its way into the general vocabulary and is widely misused and misunderstood; but in thermodynamics it has a precise and unambiguous meaning. For example, a specific quantity of a working fluid in a heat engine has an entropy that depends on its temperature and volume. Entropy is added when heat is added, in a quantity proportional to the temperature. In an adiabatic process, entropy remains the same. But the bottom line is that taking the sum of entropy for all the parts of the engine, including the hot bucket and the cold bucket if the process is reversible, the entropy stays the same; otherwise, it must increase.

In 1948 a communications engineer at the fabled Bell Labs published a paper in two parts in the July and October numbers of the *Bell System Technical Journal*. His name was Claude Shannon, and this paper was titled "A Mathematical Theory of Communication." It was among the greatest pieces of scientific analysis of the twentieth century, easily rivaling Carnot's establishment of thermodynamics. The field Shannon invented is now called information theory. Among other things, he established the "bit" as the basic unit of information. He showed that any information-bearing signal could be converted to bits, transmitted, and converted back again. And what's more, he proved that this was the most efficient way of transmitting information. We take the notion that anything can be converted into bits for granted nowadays, but it was an unheard-of idea at the time. (By the way, after the breakup of the Bell system, the part of Bell Labs that remained with AT&T was named Shannon Labs in his honor.)

One of the things that Shannon did in information theory was to notice that the mathematics of what happened in information channels looked a lot like the mathematics of thermodynamics, so much so that he called one of the measures information entropy. Shannon was smart: it turns out that information is exactly what thermodynamic entropy actually measures.

One of the most puzzling conundrums of science over the past 150 years or so is the notion of Maxwell's Demon. James Clerk Maxwell, the same one who integrated the physics of electricity and magnetism into the elegant set of equations that bears his name, also wrote a book about thermodynamics, and in it he describes a thought experiment:

> One of the best established facts in thermodynamics is that it is impossible in a system enclosed in an envelope which permits neither change of volume nor passage of heat, and in which both the temperature and the pressure are everywhere the same, to produce any inequality of temperature or of pressure without the expenditure of work. This is the second law of thermodynamics, and it is undoubtedly true as long as we can deal with bodies only in mass, and have no power of perceiving or handling the separate molecules of which they are made up. But if we conceive a being whose faculties are so

sharpened that he can follow every molecule in its course, such a being, whose attributes are still as essentially finite as our own, would be able to do what is at present impossible to us. For we have seen that the molecules in a vessel full of air at uniform temperature are moving with velocities by no means uniform, though the mean velocity of any great number of them, arbitrarily selected, is almost exactly uniform. Now let us suppose that such a vessel is divided into two portions, A and B, by a division in which there is a small hole, and that a being, who can see the individual molecules, opens and closes this hole, so as to allow only the swifter molecules to pass from A to B, and only the slower ones to pass from B to A. He will thus, without expenditure of work, raise the temperature of B and lower that of A, in contradiction to the second law of thermodynamics.[1]

The reason Maxwell's Demon posed a serious dilemma is that he appeared to be able to break the second law of thermodynamics. Maxwell gives a good account of it in the passage quoted; today we would be more likely to say that he lowers the entropy in his "vessel" but doesn't raise it anywhere else.

Maxwell's Demon offended three generations of physicists before his façade was cracked by information theory. The first reasonable assault was to note that information theory could be used to characterize the "channel" of information flow, which involved his seeing the molecules, and that the information entropy he gained by recognizing fast and slow molecules just balanced the thermodynamic entropy he "magically" removed from the system. This was a lot better than nothing, but it wasn't airtight—what if the demon used something other than sight to recognize atoms?

Suppose, for example, you gave the demon a cheat sheet that tells him what times to hold the door open and when to close it. It could be a player piano–type roll, so that the demon is just a simple mechanism. Nothing supernatural at all. And yet it would work. Now the demon has no contact with the outside world at all. You, of course, had to obtain the information from the molecules when you made up the sheet, but the demon is completely cut off during the process.

Here's the fix: noting that entropy is a measure of information, the cheat sheet has enough information on it to square up the apparent violation of the second law. The demon can separate hot and cold without violating thermodynamics, but he can't erase the sheet. With the sheet still there, the whole process is reversible; in theory, anyway, you can run all the atoms backward and have him read the sheet backward, putting all the molecules back where they came from.

In other words, the entropy represented by the information on the cheat sheet is just as real as the classical physical form in the vessel—so as long as the demon doesn't erase the sheet, he hasn't really reduced the entropy.[2]

What the second law says to modern, information-theory-aware physics, then, is that information cannot be erased. It even tells you exactly how much heat at any given temperature is equivalent to a bit of information. It turns out to be in the ball park of the heat energy of a single atom at that temperature. A simple way to understand why heat can be used to store information is that hotter atoms move faster, and thus require bigger numbers—more bits—to describe their speeds. Thus more information has been "stored in" an atom with a higher speed. You can also increase entropy by making the system physically larger. This stores more information by allowing more bits in the description of an atom's position.

So here we sit with computers that erase bits by the millions. The second law now tells us that this comes with an unavoidable price: you may think you're erasing bits, but at the molecular level you're just converting them to information in heat form.

For today's computers, this is not a problem: the amount of heat they dissipate is so much bigger than the thermodynamic limit that the difference isn't noticeable. But the computers of tomorrow, with switching elements of molecular size, will find that the limit is a problem. Not one that can't be gotten around, just one that will need to be addressed. We can first build computers to erase as few bits as possible and then simply put cooling systems on to them.

Surprisingly enough, the "can't erase bits" formulation of the second law affects machines besides computers. A simple example is a common door latch for an interior door. The bolt protruding from the door is beveled and spring loaded. When you push the door closed, the force of closing pushes the bolt back, and then it pops into place in the strikeplate hole. The process is irreversible—you can't open the door simply by pulling; you have to turn the knob.

That irreversibility erases information. If the door were a swinging door with no friction, you could tell how fast the door had been swinging a minute ago by how fast it's swinging now. The latch erases that: once it's stopped, you can't tell what it was doing before.

Notice that friction erases the same information; but we already knew that friction dissipates heat.

If you could build a one-way mechanism like a doorlatch at the very lowest of molecular levels, you could convert the random vibrations that constitute heat into coordinated motion and force, that is, useful work. But the second law, in its information theory form, stands in your way. For every bit your mechanism forgets, it has to dissipate just exactly as much heat as the amount of energy it could have gained from heat by its irreversible action.

You could, of course, build a gadget that looked like a door latch at the molecular scale. But if the spring were weak enough that it could be closed by thermal vibrations, the very same thermal vibrations would happily open it, too: it wouldn't be irreversible.

Bottom line: nanomachines cannot act as Maxwell's Demon to break the second law of thermodynamics. They cannot turn raw heat into useful work. Of course, they can turn a temperature *difference* into work, just as a macroscopic heat engine can.

As an aside, it seems likely that the mechanical losses such as friction can be kept to a minimum, so that relatively small temperature differences can be used effectively. For example, a human body might produce 200 watts of heat at about 100 degrees Fahrenheit and dissipate it into a room temperature environment of 70. Thermodynamics says that only about 5% of that can be recovered as useful work—10 watts. It's hard to imagine a macroscopic machine that could harvest that—you can lose 5% of your power in each gear or bearing in small mechanicals. Nanomachines might be able to do it, though.

THE GREAT REVOLUTION

It is ironic that much of what we know of the theories of Democritus comes to us through the works of Aristotle, who disagreed vehemently with his physical theories. Although not the first, Democritus was the most thoroughgoing, systematic, and clearly the foremost proponent in his day of the notion that matter was composed of atoms.

Chemists in the nineteenth century, notably John Dalton, revived the idea because of an interesting property that kept cropping up in chemical reactions. If you ignite 2 grams of hydrogen in 16 grams of oxygen, you'll get 18 grams of water. But if you use 3 grams of hydrogen, you'll get the same amount of water and have a gram of hydrogen left over. Chemicals react in definite proportions. Chemists were forced to the conclusion that matter, from the densest metals to the most tenuous of gasses, came in discrete, indivisible chunks.

It took the physicists some time to catch up. It wasn't until the twentieth century that atoms were fully accepted physical theory. There was a time when some physicists believed that discrete electrons were embedded in otherwise continuous matter, like raisins in tapioca pudding. However, with the help of phenomena like Brownian motion, seeing tiny spores dance around inexplicably in water under a microscope, physicists came around and by the early 1900s were busy figuring out how atoms work, untimately giving us quantum mechanics. It's another irony of science that Einstein never won a Nobel Prize for relativity; his prize, in 1921, was for the explanation of the photoelectric effect in establishing quantum theory—that light also comes in discrete chunks (which we call photons).

So by 1948 when Shannon's information theory came out, it was well accepted that matter came in discrete chunks, and that the chunks were of a small number of types. Shannon then proceeded to show that information, too, even continuous information like the waveforms of sound, was ultimately decomposable to bits. Mind as well as matter, it seemed, was digital.

The biologists weren't far behind. Within a decade, life itself had fallen to the digital paradigm. In 1950 DNA was a chemical. In 1960 it was a reel of computer tape. Over the second half of the century, biologists worked out how things like ribosomes were numerically controlled machine tools at the molecular scale.

In 2003 digital cameras outsold analog, film-using ones.[3] By now, there's a good chance that most of the pictures you see and most of the music you hear is captured, processed, and presented digitally. The text you are reading certainly is being produced and manipulated digitally, all except for the final inked page of the physical volume you hold. The digitization of information has allowed the complete and utter transformation of our ability to handle, manipulate, and transform it. It is the digitization of matter that allows life to exist at all.[4] And the control of matter in a digital way is exactly what nanotechnology is all about.

Why, after all, should we bother to build machines at the molecular scale? Large machines work just fine, and they are a lot easier to work with. And if there is some great advantage in building small, why stop at the nanometer scale? Why not picotechnology, femtotechnology, all the way down to zeptotechnology?

The answer to both questions is that the atomic scale is where matter is digital. Atoms come in a small number of discrete types and form bonds in a small number of fairly simple ways. Atoms don't wear out, and atoms of the same kind are all exactly the same. Nature has a very good quality control. To make a part that is just the same as another in a molecular machine, you take the same kinds of atoms and connect them with the same pattern of bonds. Then, in a very strong and useful sense, the new part is exactly the same as the old.

In macroscopic machines, it has long been understood that interchangable parts are a major advantage. However, no two macroscopic parts are ever actually the same. There's always a tolerance—the engineer's term for a margin for error in size. Not only does this mean that two parts won't act exactly the same, but that no part is exactly what was designed. Thus machines aren't as smooth, quiet, or efficient as they could be.

A lot of properties of everyday matter can vary with age and wear. An atomically precise part cannot wear: either all the atoms and bonds are there that are supposed to be or it's broken. Covalent bonds between atoms in molecules don't fatigue. The same thing is what makes digital computers so much more reliable, and cheaper, than analog ones. A bit is either right or wrong, never 97.4% right. A computer can do trillions of operations in a row without an error. That's virtually unheard of from an analog machine.

We understand quite a bit about taking advantage of digital properties from computers. We can use many of the same techniques in physical machines with nanotechnology. When nanotechnology makes machines cheaper and more reliable, it will be because of the digital nature of matter at the atomic scale.

NANOCOMPUTERS

Nanoelectronics is one of the leading areas of nanotechnology research today. In fact, all the components necessary to make a nanocomputer—molecular switches and nanowires—have been demonstrated. "Demonstrated" in many cases means that a lab technician sifted through chemically generated molecules for months with very expensive equipment and found a few that worked as desired. In other words, we know they exist—which is great for a scientist, but not so good for an engineer. There is as yet no way to put them together in large, complex circuits, much less to do so economically. Once we have the nanorobot arms, though, it will be straightforward.

Most electronic switches in use today are transistors. Various molecules have been shown to function as transistors. Once we have the ability to place atoms in arbitrary arrangements, the number of different kinds of known transistors will in all likelihood go through the roof.

Other forms of switches will also work at the nanoscale, and may have uses. Electrostatic relays would work. They could have fast switching times compared to today's electronics (but not to nanoelectronics). They might find a use in power applications—turning motors and lights on and off. Note that at the nanoscale, the distinction between a transistor and a relay can get fairly fuzzy. Transistors operate by having a voltage change cause a change in the conductivity of a material. In appropriately structured molecules, conductivity can be altered by moving other molecules nearby or by bending them—phenomena that seem partway between moving a physical switch and altering a property by electric fields. Molecular switches have been demonstrated in the lab using a number of different such techniques.

Most nanomachines will have nanoelectronics built in, to control all those motors, and computers to control the controllers.

In current-day computer chips, getting all the transistors you need isn't the problem: the problem is the wires. There just isn't room enough to connect everything you want with everything else. And wires have increased resistance as they get smaller.

You want wires with low resistance for two reasons. In power applications, from transmitting power to your house across the country to providing power to all those little motors that move your robot, resistance simply wastes power as heat. Lower resistance is just more efficient. That's true in computers, too, but there's another reason. Resistance makes computers slower.

Suppose you have a computer with a 1-gigahertz clock speed. That means that the clock signal rises and falls once a nanosecond— the time it takes light to travel about a foot. But the signals inside the chip are traveling much, much less than a foot. What's happening in the computer is that signals move from switch to switch by electrons filling up or draining out a wire and the part of the switch connected to it. The fullness of a wire with electrons is its voltage. Switches don't switch until the voltage changes.

Think of a switch, a transistor, for example, as a valve turned on or off by the weight of a bucket. The bucket is filled or emptied by hoses that are controlled by other valves. The crucial time for switching is not how long it takes water to start coming into the next bucket; it's how long it takes to fill (or empty) the bucket. With nanoelectronics we can make smaller buckets; the question is how big can we make the hoses?

IT'S MECHANICAL

Electronic nanocomputers will be very fast and small compared with current electronics, but mechanical nanocomputers might be smaller still. The reason is that electrons are hard to nail down and have interesting quantum properties as the scale approaches molecular size. For example, an electron sitting in a hole it doesn't have the energy to get out of can suddenly be found in another hole some distance away. This ability to jump without having been in the intervening space is called tunneling.

The probability an electron will tunnel depends very strongly on the distance it jumps. This property is used in one of the very first scanning probes, the scanning tunneling microscope (STM). In the STM, electrons tunnel from the probe tip to the object being probed, and the distance can be measured quite accurately because the current changes drastically as the distance changes minimally.

In nanocomputers or other circuitry, we are going to have to keep wires three nanometers or so apart or else electrons will tunnel from one to the other. In other words, at the nanoscale, wires don't have to touch to cause a short circuit. So nanocomputers using conventional wires, switches, voltage, and current will have to be bigger than the sizes of their components would suggest.

We could make mechanical nanocomputers that used sound conducted along strings, like a tin can telephone, to transmit information. We could use long, strong molecules like polyethylene or buckytubes for the strings. Crossing strings could actually touch without a problem, if designed right—waves traveling in different directions can go right through each other. The strings could easily be less than a nanometer wide.

The speed of sound in diamond is over ten miles per second; in appropriately stretched strings it might be even faster. If you design your mechanical nanocomputer in a conservative mode where vibrations are allowed to die out between operations, it could run with a gigahertz clock speed. An optimized design where all the wave propagation and reflection are taken into account might manage into the hundreds of gigahertz. Small, carefully tuned "circuits" could operate at a terahertz.

STORAGE

One form of computing that is tantalizingly almost possible today is using DNA. The DNA molecule is a memory device, not unlike a tape in the tape drives you see on the pictures of the big 1960s IBM computers. The DNA in a given bacterium holds about a megabyte, and that in a human cell holds about a gigabyte. (The DNA is the same stuff in each case—it's just that the human cell holds one thousand times as much DNA as the bacterium.) For comparison, this book is about a megabyte if represented as plain ASCII text and compressed pictures. So bacteria come with an instruction manual and humans with a thousand-volume set.

The natural machinery in the cell has several mechanisms for treating the DNA as a library. It avoids the library's biggest headache, books being out when needed or not being returned, by copying off information that is needed (onto RNA) and distributing the copies. The DNA itself is copied when the cell divides. It is proofread and its errors corrected. There are mechanisms to roll the DNA up for compact storage and unroll it for reading. As part of this process, it's partially cut so one part can be twisted without twisting another, and then the cuts repaired.[5]

Unless you've just had it sterilized, there is almost certainly more data storage in the DNA of the bacteria lying around on your computer, notebook, or PDA than inside as part of the device itself. At this writing, Google indexes about four billion Web pages. If we estimate the average size of a page, with pictures, as 125 kilobytes, the total comes to about half a petabyte (500 terabytes). If stored on DNA, the whole thing would fit in a 150-micron cube, the size of a grain of fine sand, or in other words, a speck just barely big enough to see with the naked eye.

Nanomachines will be able to store information more compactly than DNA, but not all that much more compactly. Using the same basic method, a series of different but similar molecules chained together, one can use smaller molecules than the deoxyribose ones in DNA, but there's not much headroom. One constraint is likely to be the competing desideratum of reading the information faster. That IBM tape drive was a lot bigger than the tape reel itself. The really dense part of the information storage system, the tape library, had racks of tapes nestled snugly together, but had a long access latency: some human had to walk down the aisle, pick a tape out, and mount it on a machine.

Nanocomputers could have a similar scheme, with nanorobots retrieving tapes, and have latencies shorter than current-day hard disks.

As a general technological truism, memory gets bulkier as it gets faster. This will continue to be true at the nanoscale, but there's so much room at the bottom that this will not be a significant problem until expectations have risen quite a bit. The fastest memory in common use is the registers and cache that operate at processor clock speeds, currently in the low gigahertz range. These are made of the same logic circuitry used in the processor. For nanoelectronics, a bit using molecular switches, probably capable of operating well in excess of a terahertz, might occupy a 10-nanometer cube. This is one thousand times less dense than DNA, meaning that you'd need a grain of coarse sand to store the whole World Wide Web instead of a grain of fine sand. Sorry! But you'd be able to access it in the amount of time it takes light to go a tenth of an inch—for the tape reel scheme, a beam of light could go twenty miles while you wait for your data.

Neither of these times is perceptible to a human, but the difference shows up when you start adding them together, as in searching the entire database, collecting statistics, data mining, and so forth. Either form could read the database at a terabyte per second (tapes are much more efficient for sequential than random access). It'd take about ten minutes to get through the entire Web. For use as a server handing out Web pages to all comers, the electronic form could retrieve eight million pages a second. The tape form could manage only ten thousand.

The amount of information the human brain stores can be estimated at something like 100 terabytes, within shouting distance of the whole World Wide Web. The brain has a cycle time closer to the 100 microseconds of our nanotape library system than the 1 picosecond of our nanoelectronic memory, but it can search its entire contents in one cycle, rather than the ten minutes that either computer-style memory would require. The brain uses something closer to what computer engineers call associative memory. (Software engineers will often say "hardware associative" memory to distinguish it from software techniques like hashing, indices, and inverted files.)

The reason it doesn't take Google ten minutes, or more, to do a Web search for you is that it's prepared an index ahead of time. This is fine for something like the Web, which changes slowly enough that indexing is a feasibly small amount of its work, and the kind of query you can ask is restricted to matching words. A hardware associative memory is one that acts as its own index, at the same speeds that ordinary memory simply retrieves words by their addresses. Associative memory is commonly used to maintain the indices of cache in microprocessors, where it would be too slow to use software techniques, and for Internet address tables in high-end routers, and in similar time-critical applications.

If the Web were stored in nanoassociative memory, you could ask things like "What's the average temperature that's referred to in Fahrenheit on pages where 'Fahrenheit' is misspelled?" and get the result back in nanoseconds.

Today, associative hardware is limited to the most demanding applications in computing, but it is ubiquitous in the nervous system of any animal more complex than an ant. This is because nature uses nanotechnology to build her computers and can afford to throw lots of hardware at tough problems. Interpreting sensor data, in particular, is a hard computational problem. Associative computers are just plain smarter, cycle for cycle, than conventional ones. When we have nanotechnology to build our computers with, we'll have the luxury of throwing some extra hardware around and be able to build smarter computers as well as faster or more capacious ones.

REVERSIBILITY, OR CARNOT REARS HIS HEAD AGAIN

Nanocomputers, whether mechanical or electronic, will have to be designed differently from conventional ones at a logical level. The reason is reversibility and thermodynamics. Every time you erase a bit in any computer, you must dissipate some energy, the amount of which depends on the temperature. At room temperature, it's about four zettajoules. This lies right in the range of the energies that atoms have owing to their thermal motion.

So if you had a nanocomputer that had a thousand atoms per gate, and each gate erased one bit per cycle, and the computer had a 1-gigahertz clock, it would try to warm up at a rate of a million degrees per second. You could beat this by brute force if you had to: a small processor, with ten thousand gates, would dissipate only 40 nanowatts. You could simply cool it with a radiator. But cooling arrangements will complicate the design, push the computing elements apart to allow for pipes and so forth, and slow the whole system down.

It's best to avoid erasing those bits if you don't have to. In a simple, one-instruction-per-cycle computer, you don't logically have to erase more than about one word—typically thirty-two bits today—each cycle. All the others can be uncomputed rather than simply erased. This is the electronic (or mechanical) equivalent to compressing the bicycle pump until it's hot before putting it into the hot bucket, and so forth. If care is taken, most of the logical operations in the computer can be made reversible. By making the logic of most of the processor's instructions reversible as well, we should be able to design computers that erase only one or two bits per cycle instead of tens of thousands.

Writing programs in a completely reversible instruction set is a bit tricky (although quite possible). Writing with a mostly reversible instruction set is not much harder than ordinary assembly-language programming. It's quite straightforward to write a compiler that turns your higher-level language code into mostly reversible machine code without you even knowing about it. So while reversibility will matter to machine designers and compiler writers for nanocomputers, ordinary programmers, much less people who just use the software, won't know or care.

QUANTUM COMPUTERS

The reason that quantum mechanics is so expensive to simulate is that quantum systems can act as if they are doing lots of different things at the same time. The classic example is that you can fire an electron at a plate with two slits on it, and it acts as if went through both of them and interfered with itself. This is called superposition of states. The basic idea of quantum computing is a simple extension: let's build a computer that acts as if it were doing lots of different computations at the same time.

A fairly substantial research effort is attempting to do this today. It faces substantial challenges. The first is theoretical: although the quantum machine could do many calculations, you can get the results of only one of them. The electron, having gone through both slits, will hit only one spot on the screen on the other side. It'll always hit in a spot where it would have been possible to go both ways and not cancel itself out, but out of such possible spots, one is chosen seemingly at random.

Thus you can't use the quantum computer as a generally capable multiprocessor with as many processors as there are superpositions. But you can do computations that would have involved trying out a huge number of possibilities, when all you need is the one that worked. The classic example is searching for encryption keys, but many other problems can be posed in a similar way: finding the path through a maze that reaches the exit; finding the sequence of chess moves that leads to a checkmate; finding a machine design that performs a given task; and so forth.

There are practical difficulties as well. To use quantum superposition to get the multiple computations, you have to be using single quanta, such as photons or electrons, to represent your bits. Thus you have to be able to manipulate them individually. In a macroscopic lab, it's difficult to handle more than a few of these in a way where unwanted influences from the rest of the universe won't slip in and mess up your computation (a process known as decoherence). This is where nanotechnology might help: building quantum computers with more superposed bits (called qubits) and where the coherence lasts long enough to perform a useful computation.

Quantum nanocomputers will blow the doors off any classical one for the kinds of problems they can do. Because of the limitations of quantum algorithms, however—you can do a vast number of computations simultaneously but get only one of the answers—they will remain special-purpose machines.

CHAPTER 8

SELF-REPLICATION
Mechanical Motherhood

G o to a park, a forest, a farm, anywhere you can see the natural world. Look about you. Virtually everything you see—trees, bushes, grass, cows, mice, insects, birds—is a self-reproducing machine. Now back to your house, apartment, cityscape. Virtually nothing is. The works of man are distinguishable from the rest of the world by being dead: they do not grow; they do not maintain or repair themselves; they must be built.

Many years ago, I had a friend who was a graduate student from Hungary. Although an educated, traveled adult, there were many parts of the American lifestyle she hadn't seen yet. I took her for dinner to the restaurant of one of the larger and more architecturally impressive hotels in the university area. As we parked, she became agitated and I asked what the problem was. She indicated the parking deck we were in and asked if it was safe. I assured her it was, and we went on into the restaurant.

Over dinner it transpired that she'd never been in a parking deck before but had seen them in movies. In movies, parking decks are invariably associated with murders, car chases in which people barely escape killer robots, or meetings with shadowy figures who are leaking classified information.

With only one exception that I can think of, the movies, and fiction in general, have portrayed self-replicating machines with the same complete lack of attention to reality and a focus purely on the horror show aspects. (The one exception is the little gadgets in *Bat-*

teries Not Included, and one cannot really point to that movie as a paragon of level-headed realism, either.)

Nanotechnology can certainly create artifacts that would give houses, cars, and cities some of the better attributes of natural life, such as the ability to maintain and repair themselves, without raising even the possibility of the ludicrous scenarios of cheap horror fiction. Unfortunately, there has been enormous misrepresentation not only from Hollywood hype factories but from "serious" commentators who should have done their homework but didn't. So let us see if we can dissipate some of the smoke and shed a little light on the subject.

Some quarter of a million years ago (or more), humans mastered our first self-reproducing technology: fire. Since then, most people have needed to know enough about fire to control it. You needed to know how to start one, maintain it at an even rate for cooking and heating, and not let it spread to engulf clothing and buildings.

Traditional fire skills are still taught by the likes of the Boy Scouts. Fire requires three things: fuel, air, and heat. Provide all three to start one; deny any one to put it out. By means of this relatively simple recipe, people have lived in wooden houses and burned wood in fireplaces for millennia. The vast majority of humans are capable of doing this with perfect safety.

Any self-reproducing technology will have a similar high-level logic. It will require raw materials, energy, and a controlling pattern. All of these will be necessary to start the process, and the removal of any one will be sufficient to stop it.

Most of us still use fire; but we don't even need to know much about it anymore. There's a fire in your furnace, one in your oven, one in your gas grill, and hundreds of fires are lit every second in the cylinders of your car engine. Machinery starts and stops these fires in carefully designed, completely controlled conditions, and in many cases you don't even know they are there.

Our second self-reproducing technology was agriculture, coming about ten thousand years ago. With this technology we reshaped the face of the Earth. Farmers labor mightily to provide the self-reproducing plants with the right environment. They do not pick the plants that reproduce most easily, but those with the most useful produce, which are often much harder to grow than the weeds.

Note in fact that people have used self-reproducing agents for centuries in applications where no one knew that actual self-reproducing agents were present: yeast in wine and bread making, and bacteria in making cheese. Plenty of people have fireplaces; make bread, wine, and/or cheese; grow plants; or raise animals for fun today. Reproduction is a wonderful process but it is not difficult to control, nor dangerous in and of itself. It isn't even all that mysterious.

By the middle of the twentieth century, scientists were beginning to realize that the functions of life were completely mechanistic and that this applied to reproduction as well. One of the great minds of the period, John von Neumann, whose name is strongly associated with the development of the computer, worked out a rigorous structure for self-reproducing automata, as he called them.

The basic idea is that you have automated production machinery, which can take raw material and produce more machinery. The machinery has instructions, which in the 1950s was typically a punched paper tape, that guide it in the making of specific parts. A self-reproducing machine would be a kind of machine shop that would be able to produce all the parts of all the machines. It would include a sort of automated assembly line, a controller that read the paper tape to direct all the machines, a (long) paper tape that directed the machines to build another shop just like the first, and a gadget to copy the tape.

Remarkably, not too long after von Neumann worked this out, the biologists who were studying the inner workings of cells figured out that at a high-level description, the way cells reproduced was just like the von Neumann scheme. DNA was the tape, and there were specific gadgets to build parts and to copy DNA.

ON THE FACTORY FLOOR

At a lower level of description, of course, what goes on in a cell doesn't resemble an assembly line, automated or not. In particular, the parts that a cell produces, protein molecules, are not picked off the production machine and carefully conveyed to meet the next part in the process. Instead they float loose to bump randomly into other parts until they meet one they click together with, in a process called self-assembly.

In a nanomachine, as in a human-sized factory, parts are never set loose to drift around. They are moved by robotic arms, shuttles, or conveyor belts from spot to predetermined spot and are assembled by mechanical force. This has two big advantages: things move much faster, going directly to the spot where they're needed rather than having to do a walkabout; and parts can be designed for their function alone, instead of having to be designed for their function and also to pop together by themselves. Designing for self-assembly is a major constraint: not only must the parts that are supposed to fit together do so, but parts that aren't intended to go together had better not fit. In a regular factory, if tab A would fit into slot C, where it doesn't belong, as well as slot B, where it does, it's not a problem; you simply don't put it into slot C.

The nanomachine is a lot faster and simpler in some ways than the machinery of the cell. On the other hand, it has a big disadvantage: it can't evolve. Cells work the way they do in part because it's easy to make small changes that leave the whole thing working. Suppose you want to add a part to an existing machine. All you need to do is start producing it. If you get the shape right, to match the existing machine, the part will bump into the machine, latch on, and start working. In an assembly line context, you need to design not only the new part, but the conveyance equipment, the assembly machinery, the process flow, and the control mechanism or software.

Now remember that in evolution, 999 changes out of 1,000 are for the worse, and die out. If you try to build a new part and get it wrong, it simply doesn't fit and nothing much happens. But in the factory model, to have any effect at all, you have to do a whole bunch of changes at the same time and get all of them right the first time.

Imagine an old-fashioned, open-air bazaar or marketplace. Instead of having fixed stalls, all the vendors have pushcarts and are constantly wandering around, buying and selling with the other vendors they bump into. Suppose there is an omelette maker. She bumps into sellers of charcoal, eggs, butter, and paper plates (all bought one at a time), at random. When she gets all the necessities, she makes an omelette and is ready to sell it to the next omelette customer she meets.

Now if you wanted to change things a bit, you could throw a few cheese dealers into the bazaar. After a while, you might reasonably expect some cheese omelettes to start showing up, as the omelette maker bumped into the new vendors at the right stage in the process.

Now imagine a factory instead. It's a building where people can't wander around. Trucks with the various ingredients come to specifically designated loading docks, one for eggs, one for butter, and so on. Inside, there's a specially equipped room for each operation, for example, one where eggs are broken, one where pans are buttered, one where half-cooked omelettes are flipped. In each room is an expert who does only that one operation. He gets exactly the inputs he needs out of holes in the wall, without looking at them because they are guaranteed to be exactly the same every time. He does his operation without looking, as well, and passes the result out a hole in another wall.

Now to change this setup to make cheese omelettes, you've got a major problem. To begin with, you have to build a whole new loading dock for cheese. Then you have to build a new room to stir the cheese in between the egg cracking and the cooking. But there's no room! The existing rooms are right next to each other. So you have to tear down half the factory and rebuild it with a new room and dock and so forth in the middle.

A cheese vendor might make a wrong turn and end up in your bazaar by accident, and then keep coming back because it is profitable. Whole, working, factory redesigns and reconstructions just don't happen by accident. Coordinated, integrated nanomachine systems with deterministic mechanical materials transfer can be much, much faster and more efficient than cells at doing some one specific pre-designed thing. They pay for this by giving up the complex, adaptive capabilities necessary for evolution. They can be adaptable if they are predesigned to handle more than one situation or process, but for every extra capability they would become less efficient. They simply cannot evolve by themselves.

This is one point the fiction writers either fail to understand or deliberately ignore. They take some properties of living things, like their adaptability, and some properties of mechanical systems, like process speed, and conjure up imaginary monsters that have both. But in the real world you have to give up one to get the other: it's a trade-off.

REPRODUCTION VERSUS REPLICATION

Researchers in the field, following the lead of Moshe Sipper at the Swiss Federal Institute, refer to systems that make copies of themselves as "replicating" rather than "reproducing." The difference is that reproduction involves making, on purpose, imperfect copies for the purposes of evolution, whereas replication is the simpler mechanical process of making an exact copy. As we have seen, living cells pay a lot for the capabilities they need to evolve.

Another thing that seems fairly necessary for evolution, but not simple replication, is sex. In sex, the genetic information describing a given organism comes from two parents, four grandparents, eight great-grandparents, and so forth. Not too many generations back and you're getting information from most of the population of your species. That means that any beneficial mutation from any member has a chance to reach you. With nonsexual reproduction, the mutation would have had to have happened to the single individual that was your ancestor. (Note that even bacteria trade DNA and are "sexual" for purposes of this analysis.)

For simple replication, sex would be an unnecessary, time-consuming complexification. Even for humans, it seems to be fairly complex and time consuming (although somehow we don't seem to mind). Sex is one more part of the price we pay for evolution. Nanomachines wouldn't need it and won't have it.

Now that we're clear on the difference between biological reproduction and mechanical replication, let's concentrate on the latter. Nature abounds with self-reproducing organisms, but where are the self-replicating machines? Are they even possible?

Even discarding reproduction, replication covers a vast range of different schemes. Von Neumann's kinematic system replicator took its inspiration from bacteria or yeast floating in a solution of nutrients. Each replicator was a little robot that floated in a soup of parts, and each time it bumped into one, it checked to see if that was the one it needed next for the partly built robot it was assembling. This corresponds to what we're calling a Stage III nanosystem, although von Neumann envisioned the parts as being macroscopic.

Here's a thought experiment: Suppose we have a warehouse full of toy robots, which can roll around on wheels but have no arms or other manipulation capabilities. Suppose each robot has a big ON button protruding from its back. As we begin, all but one are turned off. The one that's on rolls around aimlessly, and soon bumps into another's button, turning it on. Now two are turned on, and soon each will turn another on, and so forth.

Now some people would claim that these are self-replicating

robots. It's just that the raw material they use for replication is "off" robots. OK, you say, that's silly; but what if instead of just bumping a switch, the robots had to pick up a battery and install it in the target robot? How about installing a battery and screwing on an arm? An arm to a turret, the turret to a base, and put wheels on hubs?

The problem is in some sense just semantic, but it can help illuminate some questions that are not. In the first case, one hesitates to say the robots are replicating, since robots aren't being created, merely turned on. The further down the list we go, the more justification we have to claim that a robot wasn't there to begin with.

One thing to note about the first, on-switch, example is that although the robots aren't replicating, some replicating process is nevertheless present. Although it requires complex inputs and makes minimal changes, robotic activity is replicating. Another thing is that the process is glaringly obviously possible.

As we move down the list to the point where the robots are putting together parts like gears, bearings, and shafts, almost everybody would agree that a new robot is being created. This level of self-replication is also definitely, if not as obviously, possible, because it's been done in robot factories. It's more efficient, and thus more common, to have the parts put together by machinery much more specialized than the robot being built, but it has been done both ways.

Let's go back to the second example, where you have to screw on an arm and install a battery. Imagine that to do both, a robot has to use a battery-shaped gripper, install the battery, change grippers, change positions, install a screwdriver on its arm, screw on the new robot's arm, and then move back and change tools for the next incoming robot. Clearly, two robots working together, not needing to change positions or tools, could do the work more than twice as fast.

You don't need to read any further in economics than Adam Smith's original *Wealth of Nations* to hit the idea that specialization is a major source of efficiency in manufacturing. This is as true today as it was in 1776. In some sense it's more true, since Smith was talking about specialization of craftsmen and tools, and today we can make completely different machines specialized for wildly different operations.

Consider an ordinary printer as attached to your computer. It can produce a page of simple black and white text in a minute (less if it's a laser printer). You could produce a page like that with pens, rulers, and a lot of practice in what, an hour? The printer is much, much simpler than you are. You can do a vast multitude of things. The printer can do only one thing, but it does it much faster and more efficiently than you can.

The bottom line: no manufacturing system that can avoid it will use general-purpose machines where specialized ones will do. It would have been possible, using roughly the same level of technology as the printer, to have a robot arm that would write out your pages with a pen, and play chess with you, and dust your monitor as well. It would still be orders of magnitude slower than the printer.

What about replication, though? Specialized machines can't replicate themselves. Not on an individual basis, anyway. But taken as a whole, they can. Every single machine in any factory anywhere was made, in parts, by other machines in factories somewhere. The system as a whole can make any part of itself. The system as a whole can extend or repair itself even though it does not contain a single machine that can replicate on an individual basis. Such systems are called autogenous.

The more diverse the system is—the more completely specialized—the more efficient it is. Nanosystems will be able to pack millions of gears, levers, pulleys, and so forth into the volume of a human cell—plenty of room for specialization in any system big enough to see. This means autogenous systems of coordinated machines, not single replicating ones. Single replicating systems, while technically possible, would be like using robot arms holding pens instead of printers.

ASSEMBLERS

Drexler, in *Engines of Creation*, introduced the idea of molecular manufacturing by way of a device called an assembler, which was a microscopic, self-replicating, free-floating machine that could use its molecular building capabilities to build products directly after it had built enough copies of itself to make the work go fast.[1]

Assemblers are a conceptual blend of von Neumann's kinematic replicators and existing unicellular life like bacteria and yeast. We do, and have for centuries, make products like cheese and wine by growing hordes of microorganisms in vats. Thus as an explanatory device, the assembler is a straightforward extension and Drexler didn't need to expound on the morass of detail and architecture necessary for a more efficient system.

Besides exposition, the assembler has one more important role: as a bootstrap. While not as efficient as a printer at printing, a robot arm can build a printer (from the appropriate parts), while a printer can't

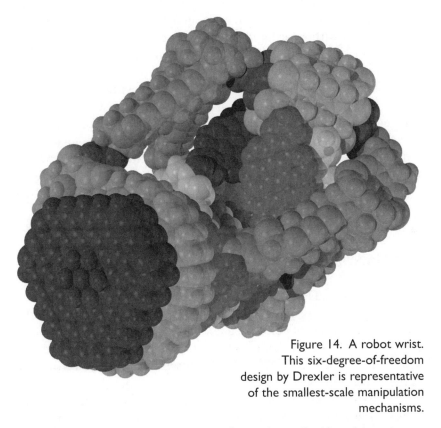

Figure 14. A robot wrist. This six-degree-of-freedom design by Drexler is representative of the smallest-scale manipulation mechanisms.

build either a robot arm or even another printer. So if you're trying to get started, you'd better build that robot arm first.

Once you get going, you'll have the equivalent of printers, specialized machines that make parts, but you'll still need the robot arms to move the parts around and put them together. Unlike the arms in an assembler, which are precise enough to make sure chemical reactions happen in the right place (i.e., better than an tenth of a nanometer precision), the arms in a parts assembly robot need only

be precise enough to put parts together. Parts can be made with bev-
elled ends, guide flanges, and so forth so that a much looser precision, maybe even a nanometer, will still get everything together
exactly.

The reason to build less precise robots is that they can move faster
and have a longer reach. You can try this for yourself. Pick up a pen
and try writing at arm's length without touching the heel of your hand
to the paper, as if you were writing on a blackboard. You'll be a lot
sloppier than with your hand touching. However, you can easily and
quickly move the pen a few feet, whereas in normal writing position
you can move it only a few inches.

Molecular deposition—controlling the chemical reactions to
build things to atomic precision, also called mechanosynthesis—is like
needlepoint; parts assembly is like stacking firewood. You could build
a stack of firewood by gluing together sawdust one grain at a time, but
it'd be awful darn slow. That's another point the fiction writers get
wrong. A simple assembler that produced something with a single arm
doing serial molecular deposition would be considerably slower than
a complex factory-like system.

MECHANOSYNTHESIS

The molecular machines in your cells are atomically precise—two
copies of the same protein are atom-for-atom the same—because they
are built by a series of controlled chemical reactions. Reactions in the
cell don't happen at random,[2] as they do in a test tube or fire, but only
where the resulting molecular structure is exactly the needed next step
in the construction process.

Similarly, in a nanomachine, the process of building a part will
consist of a sequence of chemical reactions that produces an atomically precise product. Robot arms, or factory-like machines on con-
veyors, will put molecules together in such a way that only the desired
reaction can happen. The result will be a machine part that is atom-
for-atom the same as it's designed to be.

SELF-ASSISTED ASSEMBLY

The properties of matter at the molecular scale are different from the those we're used to. Part of this is simply size. Compare the flexibility of nylon or polyester fibers with hand-sized solid chunks of the same stuff, which are hard as wood. Even diamond at the molecular scale is flexible and springy. Steel is like cheese; gold is like dough. Polyethylene is like a plate of spaghetti. Liquids are like a load of gravel in a wheelbarrow going down a stairway. Gases are like a hailstorm.

Other properties depend on the atomic nature of matter. You can't add half an atom to a part; it's either there or it isn't. Molecular parts are thus bumpy, and it can be difficult or impossible to get shapes like smooth curves.

Finally, atoms are sticky. Van der Waals attraction pulls them together even when there's no chemical bond. It's this force that causes proteins in cells to click together when their shapes match (along with some more complex interactions with water that have a similar effect).

The van der Waals force is the bane of micromachinery. Together with the fact that micromachines don't have completely smooth surfaces, the result is called stiction. It's as if all the parts were made of strong magnets coated with sandpaper. They don't slide very easily. Van der Waals force is even felt in the macroscale: it's what holds geckos' feet to the wall. The reason that all objects aren't sticky at human scale is that surfaces of macroscopic objects never match at the atomic scale. (The gecko's foot has a remarkable formfitting structure that allows it to match the surface it's touching.)

Surfaces that are smooth to atomic precision will slide, though, despite clinging to each other. So we can make use of the attraction in novel ways. For example, a peg that just fits a hole will be sucked in and will stay put if there's more than a square nanometer or so of surface touching. Thus our parts-assembly robot arm doesn't have to be very precise at all. Imagine putting a key into a keyhole in the dark. Your hand wanders around, poking at random. At last the key hits the hole and slides in. Your hand doesn't have to be as precise as the keyhole fit; it just has to be precise enough not to put the key in the wrong keyhole. If you had built the keyhole with a funnel-shaped flange, the whole thing would be fairly easy.

The van der Waals force with properly shaped parts can act like the flange. Indeed it can do more: if you have a screw and a perfectly matching threaded hole, the screw will screw itself in without your having to twist it. Just get the end to the right place. The system of design and construction that uses these properties is called self-assisted assembly.

Self-assisted assembly does impose some constraints on parts design but not nearly as much as full-fledged self-assembly. For example, you can still use standard designs for fasteners and other interfaces. The arm has enough precision to keep from putting the wrong one in the wrong place, or at the wrong time. In a full self-assembly regime, as in cells, any two flat surfaces will stick together! With self-assisted assembly, we can have all the flat surfaces we like, no problem.

So in a typical nanomanufacturing system, you'd expect to find a lot of specialized parts-making devices, producing such things as beams, casings, fasteners, diamond cable, shafts, gears, bearings, motors, wire, and computers. Then there'd be some more general-purpose devices that could make custom parts that didn't get called for as often. Similarly, there'd be fast assembly lines to put together parts into commonly used subunits, such as digitally controlled positioners, and more general-purpose robots to put together less commonly called-for assemblies. You'd expect this kind of hybrid architecture not only at the molecular level but all the way up to the size of the output products.

TEA, EARL GREY, HOT

Let's imagine the home synthesizer (personal manufacturing system) of, say, 2050. It sits on your counter and looks a lot like a microwave. It's connected to your home network, so you program it from your computer, PDA, cell phone, whatever; or just talk to it. You select or create a design with whatever software you prefer; probably most people would use an interface not unlike a Web-based catalog. We'd expect many more options for customization, though, such as ways for clothing to be sized automatically for an exact fit.

The synthesizer might be connected to the water and gas pipes and have an "odd elements" cartridge that you replaced from time to time. Then you would just tell the computer what you want and wait for the synthesizer to chime and unlock its door. The time taken would be seconds for small, simple items to hours for large, complex ones that involved a lot of nonstandard constructions. By the way, a synthesizer that used natural gas (and air) as raw materials wouldn't use power and water; it would produce them as by-products, pumping them back into the distribution network for use elsewhere. (The exception would be drinks, where more water would be used than produced.)

The synthesizer would be able to produce apples, baseballs, cups of coffee, diamonds, eggs (for eating, not hatching), folding furniture, gadgets and gizmos galore, headphones, ice cream, jackets, knives (with diamond, not metal, blades), lights, money (paper), nanosuits, office supplies, PDAs, queens (and other chess pieces), robots, synthesizers, tennis racquets (that folded to pocket size when not in use), Utility Fog, vehicles (some assembly required), watches, xylophones (saw that one coming, did you?), yarn, and zircons.

Well, actually, not zircons, since they are zirconium silicate and there's virtually no other household use for zirconium, so it would be unlikely to be available in a consumer machine. In fact, it's easy to imagine that economy models might not even handle metals, except for trace amounts in food molecules. Almost certainly a home synthesizer wouldn't be capable of making living creatures, or even eggs that could develop into them.

The reason is that the home synthesizer couldn't come close to creating any possible arrangement of atoms. It would cover a (tiny) subset sufficient for a large range of consumer products. Another subset with a different background chemistry, namely, fatty acids, sugars, starches, fiber, and proteins, could be used to make foodstuffs. The cells of your synthesized steak might be stuffed with a selection of proteins known to be safe and taste good, but not come close to the full set of molecular machines necessary to function.

Today's paper money would be easily synthesized unless the machine were specifically built not to do it. This is a problem we've seen already, as people photocopied money and printed it from their computers. The official scrip has changed form, albeit slowly, to keep ahead of the counterfeiters, and could easily continue to do so. One simple method would be to incorporate some otherwise-useless element in the official stuff and prohibit home synthesizers from having it. Gold is probably too useful; zirconium maybe?

THE SORCERER'S APPRENTICE

Since the synthesizer can make another synthesizer, isn't it a self-replicating machine? Yes and no—the term is ambiguous. It makes copies *of itself*, but it doesn't make copies *by itself*. When you make a new synthesizer, some person would still have to take it out, set it up, connect it to gas and water, plug it in, and program it before a third-generation one could appear. Thus while the replicator is autogenous, like a blacksmith's tools are, it isn't self-replicating. Thus synthesizers could have the advantages of self-replication, including very low cost, while completely avoiding any gray goo (runaway replicator) possibilities.

Even so, the availability of virtually free home manufacturing has been prophesied by some to imply the breakdown of the economic system. A classic in this vein is the *Venus Equilateral* series by George O. Smith, a radio engineer in the 1940s. Who will work when they can get everything they need from their machine?

The question can be answered by asking its opposite: who needs to work if everyone can get what they need from the machine? Even if people wanted to work, what can they do?

They can invent new stuff to be produced by the synthesizers, for one thing. This doesn't have to be rocket science, it can just be new styles in clothing. They can maintain the infrastructure that brings the water, gas, and power to the machines. They can do sports, arts, entertainment, politics, law, develop software, discover science, even write books about nanotechnology. Only 30% of our current economy is manufacturing or farming. The rest is information handling and services.

There will always be things (like land, historical objects, and people's attention) that can't be synthesized. People will work and trade to get them. But with synthesizers, the amount they will have to work to have a comfortable life will be much smaller than now.

The major problem with synthesizers is going to be annoyances we're already familiar with today on our computers. That spam offering to sell you X-rated pictures will offer to make sex toys for you in the privacy of your own home. The virus that takes over your computer and scrambles your files could poison your food instead.

Companies will offer free designs for things that work for a month and then fall apart, just as demo versions of software do now. Then they'll want a monthly fee for the working version instead of selling you a real item once and for all. They will try to sell you objects that sound useful at first but for which you have to purchase numerous expensive options for full functionality.

In some sense, it's just as well we have these afflictions now when they only affect our computers. If we met them first in physical products with the unpreparedness of early computer users, significant mischief could be done. Instead, we're going in forewarned. Are we smart enough to be thereby forearmed?

CHAPTER 9

FOOD, CLOTHING, AND SHELTER
A Day in Your Life in the Age of Nanotech

E veryday life in the age of nanotechnology could be wildly different from what it is now, but it doesn't have to be. In the early twentieth century, people were anxious to exchange outhouses for indoor plumbing, and oil lamps and candles for electric lights. Homemaking drudgery was ameliorated by everything from stoves you didn't have to build a wood fire in to washing machines. Since then we got TVs, microwaves, and Roomba the robot vacuum cleaner.

In fact, modern middle-class American life is probably too easy in a purely physical sense. We don't get the exercise our bodies need for optimal health. What use is it to add more laborsaving devices, if it only means you have to spend more time on an exercise device?

One of the classic blunders of prediction in early science fiction was the notion of the food pill. It would be a major time saver, avoiding all the folderol about cooking and serving and cleaning up— or even having to build dining rooms in the first place. Just take three square pills a day.

The problem, of course, is that most people like to eat. It's a social occasion, a significant chunk of why people enjoy life. (Early science fiction was fairly puritanical, or it might have gone on to suggest that babies should be made in factories so that people wouldn't have to bother with sex.) Even people with busy schedules and little time to spare welcome mealtimes as an excuse to take a break.

Some people like cooking, but quite a lot of the food eaten today is processed and packaged for convenience. Many who do cook do so

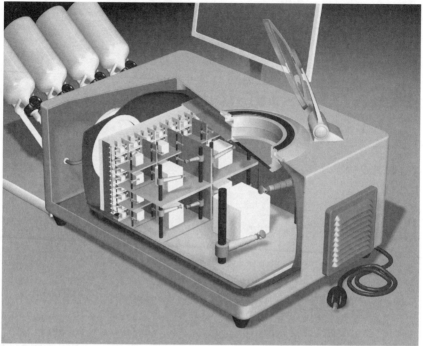

Figure 15. A home synthesizer. Also called a nanofactory or personal manu-facturing system, the synthesizer will be a universal appliance, capable of pro-ducing most household items. From a design by K. Eric Drexler, courtesy of John Burch, Lizard Fire Studios (www.lizardfire.com).

for economic reasons or because fresh, home-cooked food tastes better and is more nutritious than processed stuff.

Nanotech synthesizers in factories, then in restaurants, and finally in homes, will improve the situation. Modern processed foods suffer from two constraints that reduce quality. First, they are generally processed in central locations and then shipped and stored (and the same is true of a lot of produce). This means that what you eat isn't fresh and has been optimized for handling and storage: picked green, inundated with preservatives, bred (or genetically engineered) for shelf life instead of flavor.

The second constraint is simply cost. The reason so much of food is sugar and starch these days is that sugar and starch are cheap compared to fats and protein. (And often instead of natural fats you'll get the cheaper, but dietetically harmful, trans fats.) A surprising percentage of what goes into processed foods is right out of a chemical reactor.

Molecular synthesis will be able to make foods that are considerably closer to natural ones than current processed products. You can eat meats no animals were killed to obtain and crops no wildlife habitat was displaced to grow. No release of genetically modified organisms into the environment will be necessary to include as much of whatever vitamins and nutrients you need into whatever you like to eat. And the foods will be synthesized fresh just before being cooked or eaten, with no need even for a refrigerator. Indeed they could simply be synthesized cooked, with no need for a stove.[1]

The dishes could be synthesized along with the food, and then simply dropped dirty into the recycler. Of course, with nanotechnology, you could make solid diamond plates that would come clean with the merest wave under the faucet, but the long-term trends are clear: manufactured items get cheaper, and space gets more expensive (on Earth, anyway). So the typical domestic arrangements will tend toward making things when and where they are needed, and recycling rather than storage.

CLOTHING

His grey flannel suit fitted him with a staggering perfection, the whiteness of his shirt was dazzling, his tie shamed the rainbow. His soft felt hat appeared to be having its first outing since it left Bond Street. His chamois gloves were clearly being shown to the world for the first time. On his left wrist was a gold watch, and he carried a gold-mounted ebony walking-stick.

Everything, you understand, quietly but unmistakably of the very best, and worn with that unique air of careless elegance which others might attempt to emulate, but which only the Saint could achieve in all its glory. . . .
—Leslie Charteris, *The First Saint Omnibus*

What about clothing? The near-term products from current research include nanofibers, such as buckytubes, that are very strong. Strength of the cloth isn't enormously important in clothing, but the feel is.

Silk is an extremely strong material. It's as strong as fancy aircraft aluminum alloy, five times as strong as ordinary aluminum of the same size—or thirteen times as strong by weight. It comes in monofilaments about 5 microns in diameter, one-twentieth the width and one–four hundredth the area of a human hair. One fiber, completely invisible to the human eye, could support more than a gram (a large pill, for example).

What this strength does for clothing is that a silk thread can be quite strong enough for wearability and yet still be extremely fine. This means it can be extremely light and flexible, which are the properties of silk cloth we like. (The smoothness of feel, for example, comes from the fact that the silk threads are small and close together.) The strongest fibers we know about in theory at present are buckytube fibers, which are about 150 times as strong as silk. That means that a buckytube garment, strong enough for all normal wear and tear and to hold all your junk in its pockets, would be so diaphanous that you'd have a hard time knowing whether you had it on or not.

Clothing is one of the very few really old human technologies. For almost all of our history, we wore clothes made from animal skins and woven plant fibers. Only in the twentieth century did that change, and then only to include synthetic fibers. Most of these still don't feel as good as, say, cotton, although they're getting better fast. (It may even be the case that we've evolved to like the feel of woven plant fiber on our skins—we've been wearing clothes that long!)

The big difference with clothing will be that your closet is a synthesizer, not a storage bin. You'll have a "magic mirror" that shows what you'll look like in whatever you're thinking of wearing. Most likely you'll be able to talk to it to select and modify the style or fit. Then it's synthesized, you put it on, wear it, and toss it into the recycler. No laundry.

The major functions of clothes are for protection from the environment, modesty, and social display. Technology, nano or otherwise, has only an indirect effect on social display. First, the clothes you wear serve to identify you with a class or group, a uniform formal or informal: T-shirts and sandals for programmers, suits and ties for businessmen, clerical collars for the First Estate. Clothing makes a statement of personal style as well as a group identification. As with many matters of style, the logic seems sometimes to run backward: a business suit has its value because of what it prevents you from doing. It thus advertises that you are a person who doesn't need to do manual labor. (This advertisment was particularly pronounced in upper-class women's fashions of the Victorian era, for example.)

It seems quite likely, then, that there will remain a market for hand-tailored clothing of classical materials for quite some time, and that the more expensive such clothes are, the better status symbols they will make. Especially since it will be easy enough to wear a nearly unnoticeable nanosuit underneath and obtain all the benefits anyway.

THE NANOSUIT

Imagine a fabric just 5 microns thick, the thickness of that invisible filament of silk. It covers your whole body, like an extra layer of skin. (There's a separate hole for each individual hair.) With about two square meters of skin to cover, it's the same weight as two teaspoons of water. To get an idea of how thin it is, take one drop of water and spread it over your hands as if washing them. To get an idea how heavy the suit will feel to wear, swallow two teaspoons of water and walk around naked.

The fabric, however, is not woven of threads. It is a machine. To get an idea how much machinery could be in it, look under the hood of your car. You'll see wires and tubes of about a quarter inch diameter, some pipes one or two inches in diameter, pulleys half an inch wide, shafts the same width, subassemblies in the six-inch range. Now go to a Wal-Mart or similar giant warehouse store. Imagine that the ceiling is eight stories high, and that the building itself covers the entire continent of Asia (or North America, twice). Now pack it full of machinery like the engine compartment of your car—that's your nanosuit, scaled up by a factor of five million.

The major function of the suit is the same as the animal skins our ancestors wore tens of thousands of years ago—to keep warm. With the nanosuit, however, there are a few advances. To make the fabric flexible yet strong, it contains two or more layers of plates that can slide past each other, like scales on a fish. Each plate is hollow and contains vacuum, and has a reflective coating to prevent radiative heat transfer. Among the machinery are plenty of reversible heat pumps. When the environment is too hot, they pump the heat away from your body (requiring an energy source); if it's too cold, they act as heat engines, picking up energy from the flow of heat from your body to the outside. (We note that most environments on Earth where the outside temperature is hotter than the human body are blessed with strong sunlight, which could be used as an energy source.)

You are now set to go play tennis outdoors in Nome, Alaska, in January (assuming you have a lighted court). If you look at a globe you'll begin to see just how much difference such a relatively simple nanotech application could make to the human race. It could easily double the livable land area and substantially ameliorate one of the major energy needs.

If you like to live on mountaintops, or under the sea, extend the nanosuit to provide oxygen. Molecular sieve oxygen concentrators exist today, and using powered machinery designed at the molecular level would make them easily wearable. For gills, you'd need a forced-water feed to the filter, since dissolved oxygen in water is at a much lower concentration than in air. On the other hand, if you cared to carry a power source, a rebreather that reconstituted the oxygen from the CO_2 you exhaled would be relatively straightforward.

Computers and other high-tech gadgets as part of clothing have been in the labs for some time (notably, the wearable computers of MIT's Media Lab). Although not integrated into clothing, the set of gadgets that a well-dressed businessperson carries has been expanding rapidly in recent decades, past the watch to the pager, cell phone, and Palm Pilot. All these functions could be built into the nanosuit, and many more besides. Although the suit will be less than a tenth the thickness of a typical human hair, you could put the equivalent of 100 billion modern-day PCs into 1% of its volume. Surely unnecessary, but it's clear that the processing power for any and all computing, communication, interface including virtual reality, and like needs would be available.

The suit could be covered with tunable optical reflectors and thus be a high-resolution video display, showing whatever the wearer desired. It could even have phased-array optics, that is, a holographic video display (there's a use for a few billion processors!), and be able, among other things, to render the wearer nearly invisible. On the other hand, the sensory capacities of a suit, if desired, could be just as much advanced over natural human ones and would probably be able to unmask an "invisible" suit fairly easily. In practice, such capabilities would probably be used mostly for expressions of personal style, including full color and motion tattoos, projection of the appearance of holes through one's body, presentation of oneself as a skeleton or a member of the opposite sex, or other personal idiosyncrasies.

LITTLE HOUSE ON THE PRAIRIE

Home ain't a place that gold can buy or get up in a minute;
Afore it's home there's got t' be a heap o' livin' in it;
Within the walls there's got t' be some babies born, and then
Right there ye've got t' bring 'em up t' women good, an' men;
And gradjerly, as time goes on, ye find ye wouldn't part
With anything they ever used—they've grown into yer heart:
The old high chairs, the playthings, too, the little shoes they wore
Ye hoard; an' if ye could ye'd keep the thumb marks on the door.
—Edgar Guest, *A Heap o' Livin'*

Although marks of wear and soiling may disappear in the self-cleaning and self-repairing structures of the future, perhaps their value as mementos will be replaced by video recordings, already common, and similar improvements as time goes by. Indeed it would not be difficult for a house to record everything that transpired inside. At molecular storage densities, a gigabit signal for a century could be stored in about 3 cubic millimeters, the volume of a toenail clipping.

The house itself will probably be a lot more configurable than the static structures we live in today. Objects, including furniture, will be cheap, and space will be at a premium. Alternatively, and for the same reason, houses will be put in more and more remote areas and will be responsible for the well-being of people under extreme conditions. If a vanishingly thin skinsuit can keep people warm in the Arctic, the walls of a building can easily do so. Warming and insulating houses from the cold, and cooling them in warm climates, is widespread current technology; nanotech will just make it cheaper, more effective, and less obtrusive.

Power in remote areas can be had from solar collectors. Sunlight is too diffuse for economical collection with current technology, but with nanotechnology you could do something like replace the grass on your lawn with a carpet that looked and felt like grass, never needed mowing, and acted as a solar energy collector. This would give you enough area to pick up a decent amount of power without having to build giant structures all over the place. The other current problem with solar power is storage. Nanotechnology should be able to handle that one fairly easily.[2]

Another current technology that makes quite comfortable living is the houseboat. Nanotechnology could bring the cost down and improve the reliability enough for houseboats to be a viable alternative to land-based housing. Living on the high seas would be significantly more feasible with synthesizers for most everyday needs. Depending on social and political arrangements and personal preferences, there could be single-family boats and/or cruise ship–sized floating cities.

Nanotechnology could also make living underwater a viable option. Submarine dwellings would be less vulnerable to storms than surface ones, but in good weather, much more energy is available on the surface as sunlight. The ultimate trade-offs remain to be seen.

THE TOPLESS TOWERS OF ILIUM

Christopher Marlowe referred to the towers of Ilium as topless not because they didn't have roofs, but because they were unexcelled in height.[3] But a nanoengineered tower, with adamantine structural materials, could be so tall it would literally disappear into the sky: more than ten miles or so up, most of the blue of the sky is below you. To eyes on the ground, it would fade into invisibility, even in the absence of clouds or haze. On a clear night, of course, its lights could be seen all the way up. Towers could easily be 100 kilometers (60 miles) tall; much taller than that and they begin interfering with satellites!

In 1956, Frank Lloyd Wright designed The Illinois, a mile-high tower that would accomodate one hundred thousand people. (It was never built.) Scaled up by a factor of sixty, such a tower would have a footprint of about ten square miles and have more than thirty thousand floors, and thus three hundred thousand square miles of floor space. By scaling up Wright's calculations it should be able to house twenty billion people. That wouldn't work in practice (nor would Wright's actual design), because you can't simply use apartment building densities for something that has to include city-scale infrastructure, services, and transportation. However, you could almost certainly house one hundred million people, maybe up to a billion. Thus you'd need somewhere between five and fifty such towers to house the Earth's current population.

After the first mile or so, the tower would need to be pressurized just as airliners are. I certainly wouldn't want to live in such a tower, but plenty of people live in apartment buildings now; it's a lot better than nothing. It could be made much more comfortable than current apartment living with the space versus objects trade-offs discussed above, and built-in virtual reality. You'd rarely experience the box you were actually living in.

VIRTUAL REALITY

> Yesterday upon the stair
> I met a man who wasn't there.
> He wasn't there again today.
> I wish that man would go away.
> —Hughes Mearns, "The Little Man"

Virtual reality is a term that's used today for various advances in entertainment and user interface technology, but it's all heading for the same basic capability: to make it seem, to all your senses, that you were in a different, often an imaginary, situation than you are. Today, an Omnimax theater does a good job with sight and sound. Full-fledged virtual reality will extend that in two ways.

As far as your senses are concerned, the other major sense (actually a cluster of related senses) is feel. However, the really hard part will be to allow you to interact with the simulated world, manipulate it, do something not in the script.

Virtual reality can be divided into two categories, depending on where the "world" originates. In pure virtual reality, it's simply made up, in some computer's imagination, like the world of a video game. In telepresence, it's a real place—you just don't happen to be there in actuality. You would most likely use the same interface to do both kinds; in one case it would be connected to a computer (as would other people who wished to join you in the imaginary place); in the other, it would be connected to a robot that would "be you" in the distant, real, place.

Nanotechnology will make high-fidelity android robots affordable (they are close to technically possible now). A major use of such androids will be for telepresence. There will be a wide range of applications, ranging from tourism to business meetings to scientific expeditions. It will be fairly simple to make the remote android reconfigurable to reflect the appearance of the user. It will also be fairly simple to make the appearance of the android *not* reflect the appearance of the user.

The most unbounded form of virtual reality, including telepresence, will be by nerve taps, where synthetic signals representing the signals your sensory organs would be getting from the imaginary world are injected into the appropriate sensory nerves, and the signals going through your motor nerves are intercepted and used to play your part in the simulation.

The real danger from these possibilities is that the virtual world is so rich and full of possibilities that it may be quite difficult to get people to pay any attention to the real, physical, one.

AN INTERLUDE BEHIND CLOSED DOORS

One trend that was very strong in the West during the twentieth century was the loosening of moral strictures on sexual activity. It's often claimed that this is because of the contraceptive pill, but it is quite a common phenomenon in an increasingly cosmopolitan society. Ancient Greece and Rome are examples.

As with other forms of entertainment, nanotechnology will make possible a wide variety of new sexual modes. Simplest is that nanomedicine is very likely to make contraception even easier than it is today, and for either partner. It will also substantially reduce sexually transmitted diseases. And drugs and treatments that enhance sexual performance and satisfaction are certain to appear. The immediate result could be a 1960s-like sex fad.

Another thing that nanotechnology will bring is android robots. It seems quite likely that toys very difficult to distinguish from human beings will be available, given how much people like to play games with other people. These robots would not only look and feel like humans, but would be intelligent, emotional, and come with completely individual personalities. There will be a strong debate about the morality of using these toys in ways that would be unacceptable with real humans.

Sex with real people through telepresence will be popular as well; not only will it be possible to do it in places distant from where you happen to be, but in places where, for example, it would be quite uncomfortable for a human to be naked, like a high mountain peak or at the bottom of the ocean.

The most common form of sex, though, would most likely be purely virtual. It's highly likely that the process of dating will be as accelerated as that of business meetings. Your account on the Internet dating site would come with access to virtual restaurants, parks, entertainments, and bedrooms. Instead of typing in a chatroom, you could engage in anything from a moonlight stroll to an orgy.

Because of the cleverness of interfaces, though, some of the problems of current Internet dating will remain. Certainly there will be a temptation to improve one's appearance through the interface; it will be possible to spend the night with someone and still not be sure what the person really looks like. (On the other hand, nanotech plastic surgery will be popular as well—so maybe most people will really look like they wished they did.)

Pure virtual reality opens up virtually limitless possibilities, for games of all kinds, for assignations with partners real, imaginary, or mixed. The human participants don't even have to experience the same thing; one could be having sex and the other debating politics, if the interface program were sufficiently clever.

> The final prediction we can make with confidence, if some impatience: Weightlessness will open up novel and hitherto unsuspected realms of erotica. And about time too.
> —Arthur C. Clarke, *Profiles of the Future*

CHAPTER 10

ECONOMICS
How Much Is This Going to Cost?

Fifty years ago you could get a nice car for two thousand dollars. Today, a nice car will cost something on the order of ten times as much. Of course, you get more for your money. Today's cars have better gas mileage, emit less pollutants, are quieter, handle better at high speeds, are safer in crashes, and last longer with less maintenance. You're more likely to have automatic transmission, and you may get on-demand four-wheel drive. That's not to mention air-conditioning, four-speaker CD players, power windows, and keyless entry systems.

The price difference is deceiving. In fact, inflation has reduced the value of money by more than a factor of ten since 1950. In other words, today's car costs about the same, in constant dollars, as the 1950s one. You're getting quite a bit more for your money.

If you look at computers, though, they leave cars in the dust. By now, it's an old joke what cars would be like and how much they would cost if they had kept pace with computers over the same period. Just to get some feeling of how computer technology has changed, though, remember what a typical university computer was like when I was in college in the 1970s. We had the university records in a punched-card file that formed a twenty-foot wall in the computer center. At five feet high and a yard deep, it weighed about ten tons. There were about two hundred drawers holding three thousand cards each, and each card stored one eighty-character line of text. That's less than fifty megabytes: the capacity of a stamp-sized flash chip for a camera or a USB stick today is typically five times as much.

Dollars per Megaflop

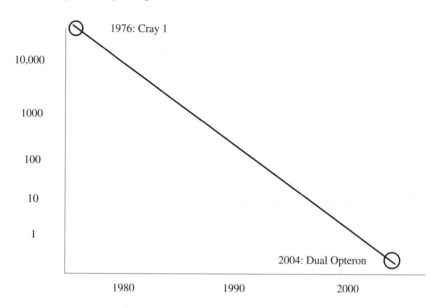

Figure 16. The cost of computing. Comparing the legendary Cray-1, at 133 megaflops and $7 million, with a current-day computer similarly configured as a scientific number cruncher, which is sixty times as powerful but costs only $3,000.

The computer itself cost $100,000 and was bought with a grant from the National Science Foundation. It had 65,536 bytes of memory. Each bit was an actual ferrite core, threaded by hand. It could process approximately 250,000 instructions per second. In other words, it would have taken ten thousand such computers, at a cost of a billion dollars, to do the the work of a good PC of today. And that's only in thirty of those fifty years.

Now we come to the really interesting question, which is in some sense at the very heart of the nanotechnology. Is it possible that over the next fifty years, the products of physical technology, motors, cars, factories, airplanes, what have you, could undergo the same kind of optimization and price reduction as computers did over the past fifty? If so, it's quite reasonable to think of owning high-tech stuff whose equivalent today would cost a billion dollars, like a large factory, an ocean liner, a fleet of aircraft, or a spaceship. You could reasonably expect to be able to carry something in your pocket whose equivalent today would weigh ten tons, like a comfortable dwelling.

The key to the price of something is the costs of its factors of production. That includes the raw materials, capital equipment, land, labor, transportation, middlemen, taxes, and so forth.

It should be pretty clear that if you have a synthesizer sitting on your counter, then land, labor, transportation, and middlemen disappear. Even if there were a tax on synthesized things, taxes tend to be levied as percentages of costs—they would remain in proportion. This leaves the raw material, which could be coal, natural gas, oil, or even wood. Delivered to your home in ton qualtities, these materials cost from 5 to 35 cents per kilogram. If you convert them directly to a one-ton car, the raw materials cost of the car would be, say, $200. (In fact, a nanotech car could weigh much less than a ton.)

Suppose a synthesizer weighs in at 10 kilograms, a dollar's worth of coal. Suppose it takes a synthesizer an hour to make another synthesizer. In other words, if I have a synthesizer, I can make 8,760 new synthesizers in a year. Since I can borrow money at 5%, if I can sell them at $2 apiece, clearing $1 over the raw materials, I can break even if the first one costs less than $175,000.

But it would be really dumb to sell the synthesizers off as I made them. After the first hour, I have two synthesizers. If I use both of them to build more, I have four after two hours, eight after three hours, and so forth. In fact, I have over sixteen million synthesizers after twenty-four hours. At 5% annual interest, the interest for one day is 0.0137%. If I can clear $1 on each new synthesizer, I break even if I borrowed $116 billion for the first synthesizer.

Of course, no one is going to lend you $100 billion to buy a synthesizer and accept the synthesizer in repayment if you're now selling them for $2. The example just shows how an autogenous technology has the mathematical capability of driving a billion-fold price reduction.

In the real world, it might work something like this. A company invests $10 billion in making the first synthesizer. They sell ten to other companies, for a billion each, recouping their initial investment. Each of these companies sells ten at $100 million each, and so forth, $10 billion total being exchanged at each stage for ten times as many machines, until a billion machines are sold for $10 apiece, no one having lost any money.

Alternatively, the inventing company could hold the autogenous technology closely, selling only products of the machines and consumer synthesizers that could make products but not more synthesizers. Home synthesizers could be built to require a payment to the company for each object produced. Handled right, they could amass a tremendous monopoly and extract enormous amounts of cash from the public.

Since synthesizers, or indeed any form of autogenous technology, could be such a huge cash cow, any company that had it would go to extreme lengths to prevent others from getting it. This is what economists call monopoly rent; you are willing to spend a large part of the difference between your monopoly price and a fair market price to protect your monopoly, even though the resources so used do no good to society at large (and indeed generally do harm). The first place a would-be technological monopolist looks is for patents.

A patent is simply and purely a grant of monopoly. Why would a supposedly enlightened government, which has laws against monopolies in other forms, grant them? The original idea was the opposite: you wanted the inventor to publish a description of the invention instead of keeping it secret. To induce him to, you offered, legally, some of the protection that he would have gotten by keeping the secret, enough to get a good head start on the competition.

It's not a bad idea, if it were done right. But it isn't.[1] Because the patent office makes money on each patent issued, it has an incentive to patent any silly thing. Suppose the invention is obvious enough that it's easier to reinvent it from scratch than to search the patent records, interpret, and adapt the record of the patented version.[2] Then the economic effect of the monopoly grant is a pure, unalloyed loss to society. Even if it were relatively costly to reinvent the gadget or whatever, that cost would have to be balanced against the social cost of the monopoly rent of the patent.

A side effect of patents is that when someone patents a gadget that turns out to be profitable, everyone else tries to invent a new one that does the same thing a different way, to get around the patent. So it's virtually certain that people will be trying to build autogenous technology using every conceivable scheme, architecture, and design. Not only will this lead to the deployment of less efficient schemes, but it will probably push the late arrivals toward designs that could be more dangerous, such as loose replicators.[3]

One other thing about patents: they do require you to publish the design of the thing you're patenting. And they do not prevent people from building the thing, only from producing and selling it. Building the invention for research purposes is explicitly allowed. Patents were, after all, intended to advance science and the useful arts.

So we'd expect there to be lots of research replicators built. Depending on the investment involved, this could be by companies or hobbyists. It's not clear whether this is good or bad—it's just something we should expect given our current legal situation.

It would be nice to think that if autogenous technology were developed by one company that produced a standard line and sold them to everyone, better control of safety and quality would be relatively easy to maintain. Unfortunately, history says differently.[4] For example, in the 1970s, American automakers let quality slide so much that Japanese carmakers were able to make large inroads into the market using quality as a selling point (by 1980, that is; earlier the opposite was true). Monopoly engenders poor quality as well as high prices.

Another monopoly rent effect is that the would-be monopolist designs its products to lock in its customers. In other words, the products are compatible only with other products of the monopolist. A classic example is in the 1960s and 1970s, when IBM was the eight-hundred-pound gorilla of the computer world, and IBM computers used the EBCDIC character code while the standard for the rest of the world was ASCII.[5] (ASCII remains the standard today.) IBM files looked like gibberish on non-IBM computers and vice versa. (With the loss of their dominant position, IBM systems are now much better team players.)

At the same time, cooperative efforts of hobbyists and amateurs have built operating system software of markedly better quality, such as Linux, which is given away for free. The economics of replication are such that synthesizers, and designs for their products, could be produced the same way. Churches and other eleemosynary institutions do not give away free software today, but synthesizers are a more direct and obvious application of the charitable instinct. Synthesize someone a meal and you feed him for a day; give him a synthesizer and you feed him for a lifetime. The necessities of life, indeed for quite a comfortable lifestyle, could be had by the general population for not much more than the price of the raw materials.

ROBINSON CRUSOE

Imagine you are on a deserted island. There is enough plant and animal life for you to support yourself working several hours a day, but you aren't living in luxury by any means. Now a crate washes up on the beach and out climbs a robot. It's designed to serve humans. You're the only one around, so it's all yours. It turns out to be stronger and faster than you are, skillful, smart, eager to help, and tireless.

First thing you know, you're in a paradise. The robot does all the work and even provides some luxuries. You can do whatever you want: relax, play music, and attempt projects you didn't have time and energy for before.

Now suppose instead ten people live on the island and ten robots show up. First thing you know, the people are bickering over who gets what. The robots are all working for some of the people doing what the others used to do. Some people are living very luxuriously and others are less well off than before the robots came. Much of the robots' efforts are bent toward keeping the less-well-offs from getting what the more-well-offs have.[6]

Economics is the science of why the second scenario is so different from the first.

Money is a remarkable invention. In a premonetary tribal setting, people traded goods but gained flexibility by remembering who their friends were, who was generous and thus deserving of generosity, and so forth. Cheaters were known by reputation and soon learned that they would have a much harder time of it if they did not make amends. The problem with such a system is that it doesn't work for communities of more than about two hundred people. The human brain can't learn and hold information at the rates and volumes necessary for an "everybody knows everybody" society much bigger than that.

Money allows for fairness in economic dealings of more or less unlimited scope. The amount of money you have, in theory, represents the amount of goods and services you (or your ancestors) have provided to others, minus that which has been provided to you. So if you have the money to buy something, you deserve to have it. In practice, it's a lot more complicated. Still, it works well enough to allow worldwide trade and to organize vast patterns of activity involving people who've never seen each other.

The alternative to money is the "command economy." In this model, some entity, be it a tribal chieftain, factory boss, people's revolutionary council, or giant superintelligent computer, simply tells everyone else what to do. This model works better than money for groups up to two hundred or so, assuming the boss is competent. At larger scales, it becomes less efficient. At a national scale, it's a disaster. When the Soviet Union broke up, various venture capitalists studied its industry with an eye to setting up new companies. Not much happened, because it was found that on the average, the products that Soviet industry produced were worth less on world markets than the raw materials used to make them.

Still, there is the problem of automation, made worse by the expanded capabilities of nanotechnology. Are we all doomed to be thrown out of our jobs and replaced by machines? Is some form of socialism necessary to avoid it?

In common automation scenarios, the company automates some process previously done by humans, throwing them out of work. People have worried about this sort of thing over most of the twentieth century. Some 90% of farmers have lost their jobs since 1900, replaced by automation.[7] Manufacturing has shrunk, not as drastically, but farming and manufacturing together today are only about 30% of the economy.[8] The rest is information handling and services. There were no computer system administrators in 1900 and many fewer restaurateurs than now. Does that mean that the trend will continue, until all goods are made by machines?

Note, too, that information and service jobs are not immune to automation. Robotics and AI (artificial intelligence) are progressing rapidly, and it's easy to project a significant proportion of automation in these fields over the next few decades, nanotechnology or not. Eminent roboticists such as Hans Moravec predict an inexorable tide of robotic capabilities washing over the successive foothills of human ability and replacing us in the economy.[9]

Just as a thought experiment, let's assume that nothing a human can do can't be done better by the machines. The human race as a whole should be in great shape. Collectively, we're in the same position as Robinson Crusoe and his supercapable robot Friday.

Even individually, we'd love to have our jobs be taken over by machines. Simple scenario: Someone gives each person a robot that can do his job.[10] The robot does the work, and the person still draws the pay. No problem! The problem arises not in the existence of the robot, but in the question of who owns it. Giving each person a synthesizer has the same effect. We've seen that it's feasible to do this with nanotechnology. People could be more truly independent than we've been since paleolithic times, when we were hunters and gatherers, able to live off the land.

DRAWBACKS

Would we be demoralized at having our effective roles being taken over? Would we feel worthless, dependent? Humans take pride, indeed base a large portion of our personal identities and self-esteem, and tend to categorize other people, in terms of what they do. He's a carpenter, she's a reporter. In a tribal setting, such stereotypes wouldn't mean as much since you'd know more about the individual. But also in a tribal setting, that pride is an integral part of making the society work without money. The respect or disrespect of one's peers is at least as strong an incentive to work as is the prospect of material gain. And in the long run, the respect of others, not any particular amount of goods and services consumed, is the key to a life well spent and satisfying in retrospect.

Crusoe was better off with his robot because the robot did things that Crusoe wanted done. Psychologists, notably Abraham Maslow, have posited a hierarchy of needs that ranges from basic physical necessities like air and food through various psychological and social desirables.[11] A salient point of the model is that the things you're interested in and worried about are at the level in the hierarchy where everything below them is satisfied, and taken for granted. But wherever in the hierarchy you are, there's always something you want. So no matter how much Crusoe's robot does, there's always something else for it to do. He could have a troop of robots, an army, a nation of robots, and he could still see things just beyond their grasp.

So with the human race as a whole. Once we have the robots, our job becomes deciding what it is we want to have done. After that, economics becomes the study of how we translate the individual desires of persons into the collective desire of the whole human race.

There is one clear, overarching danger to the whole process. It shows up in a wide variety of forms, but it can be summed up simply.

We must make sure that we are the masters of the robots, rather than their slaves. The main horror of the socialist vision is that it makes people part of a machine, rather than making machines extensions of people. The same can be said for corporations.

The devil, of course, is in the details. How do we get there, to the good vision, from here, while avoiding falling into the bad vision? To begin with, let's try to examine the good vision in more detail.

UTOPIA, LTD.

We start by giving everyone a house, a synthesizer, and an Internet connection. We'll assume that there are corps of robots that build and maintain the infrastructure. Each person, or family, is thus provided with basic physical amenities and is not obliged to work to stay alive. As an autogenous technology, synthesizers and robots will be dirt cheap in a Stage V nanotechnology; but land will get more and more expensive. Ultimately we're going to pack people like sardines in megahives or move into space, or probably, both.

People will need to produce something to trade in order to get land or other limited resources. They'll probably need to produce something and give it away in order to get other people's respect and attention. A current-day example is Weblogs, where people do the equivalent of writing a column for free, thereby getting their opinions propagated, their names mentioned, and so forth.

Virtually everything people produce will be information. The instructions for your robots are information. Entertainment produced and consumed is information. Specifications for anything to be built in your synthesizer is information.

Current-day costs of reproducing information are minimal. If you walk into a store and buy a software application in a box, it will typically cost you tens or hundreds of dollars. The actual cost of hosting a 100-megabyte file is two cents per download. With a nanoengineered, robotically maintained infrastructure this should shrink to virtually nothing. The true cost of information will be the cost to produce it divided by the number of people who want it.

Turning a ton of coal into a ton of robots is essentially a process of imposing information, structure, onto it. The bulk of the cost of a ton of robots should be the cost of the raw material. The information may well have originally cost a lot to produce, but it's even cheaper to duplicate than synthesizers. The only things standing in the way are monopoly rents and transactional inefficiencies.

Limited resources, such as land and energy on Earth and matter in space, can be allocated in a number of ways. Most simply in concept, each person gets a fixed amount. This is problematic in practice since people will trade, and there will be new people. You're constantly having to take stuff away from some people and give it to others, which cannot be done in a way that people will agree is fair. The far side of the spectrum is property, where what you have depends on what you had, plus what you did with it. New people get only what their parents decide to give them.

A hybrid scheme seems possible. Historically, the real interest rate,[12] which reflects the productivity of capital, has been 3% per year or thereabouts. In a technology as productive as nanotechnology, the real interest rate might climb substantially because of the shortening of the time that it takes a unit of capital to produce another unit of capital. Think of it this way: why should I lend you money at 3% when I could buy a synthesizer and multiply my capital a millionfold instead? To compete with that possibility, you have to offer a much higher interest rate.

Another way to look at it is that the total value of all the stuff in the economy will climb faster. That means that the government, or whoever is managing the money supply, can simply inflate to cover the difference. Prices will remain stable, since there will be more goods for the new dollars to chase. And you can give the new money out evenly as a dividend to the citizens, while allowing them to keep whatever they have, can create, or can trade for. Taxes should be unnecessary.[13]

GETTING THERE

> There is nothing more difficult to take in hand, more perilous to conduct, or more uncertain in its success, than to take the lead in the introduction of a new order of things. Because the innovator has for enemies all those who have done well under the old conditions, and lukewarm defenders in those who may do well under the new.
>
> —Niccolo Machiavelli, *The Prince*

Once upon a time, an apocryphal queen had two prisoners in her dungeon. Wanting justice to be done, but not knowing whether either was actually guilty of the crime accused, she ordered each to tell whether the other was guilty. If both maintained the other's innocence, each would get a two-year sentence. (This is referred to as cooperating.) If both maintained the other's guilt, each would serve five years. (This is called cheating.) If one prisoner called the other guilty while being called innocent, the cheater would get one year and the cooperator ten years.

From the point of view of either prisoner alone, the obvious thing to do is to cheat. He'll serve only half as much time as if he cooperated: five instead of ten if the other also cheated, one instead of two if he cooperated. But for both the prisoners combined, they'd be better off cooperating than any other combination: four total years instead of ten or more.

This little conundrum is called the Prisoner's Dilemma, and it is a sketch in miniature of a lot of economic interactions. The essence of it is a situation where the best thing to do from an individual's point of view is different from the best thing for everyone to do from the point of view of the group as a whole.

An important point to remember about the Prisoner's Dilemma is that it really is a dilemma. If it were always better for people to take the community point of view choice, we could have evolved to do it with complete selflessness, like ants. Instead we evolved to be flexible, and to be able to think about dilemmas like this. Nanotechnology won't solve it, nor will any other physical technology. Even if we were to genetically reengineer people to be selfless, we'd all be worse off. People's individual intelligence is much better at understanding what would help them, or their small group, than the nation or world at large. A large chunk of the good of the whole really is the sum of the good of the parts. The trick, and what evolution has tried to do with humans, is to strike a balance.

The major roadblocks between us and the good vision are Prisoner's Dilemmas. Governments (or corporations or individuals for that matter) don't get more efficient without a lot of pressure. Most people and corporations involved in the development of nanotechnology will be out to gain the greatest advantage for themselves, and in many cases that will mean restricting the benefits for as long as possible to get as high a price as possible, as in the discussion of synthesizers above. Indeed, corporations are required by law to act this way—they can be sued by their stockholders if they don't.

In the extreme, people (corporations, government agencies) who occupy an advantageous position in today's prenanotech world will find it in their interest to oppose any nanotech development at all. They will at least try to get the particular benefits that encroach on their current turf banned or regulated.

This monopoly rent seeking is a particularly vicious case of the Prisoner's Dilemma. If everyone in society has a monopoly, none is gaining an advantage, and all are being made poorer. If some wish to be nice, and cooperate to mutual advantage, they can do so and exclude the cheaters—in a free society. Cooperators thus have an advantage, and can prosper. Cheaters are forced to deal with other cheaters and are at a disadvantage in the long term. Thus cooperation can evolve. But in a society where cheating, that is, monopoly, is enforced by law, the only way cooperators can prosper is by operating outside the law.

Again, this is *not* a problem nanotechnology can or will solve. It's a problem that will impede nanotechnology from solving the things it could, and is another of those features of the existing world, unlikely to change, that will make nanotechnology more dangerous than it need be.

Another roadblock looms between us and nanotechnology. Many affluent Americans and Europeans are quite comfortable today, thank you. It's not politic to come right out and say "I've got mine, screw you," but an industry has already sprung up of apologists who will try to demonize any technology that seems to have the capability of drastically improving the human condition for everyone.

This being the case, the prognosticator is also on the horns of a dilemma. The majority of nanotechnology research today is being done in the United States and Europe. These are just the places where the currently affluent and powerful have the least need of it and where alarmist arguments are most likely to get a hearing. In plenty of places in the world, however, people are willing to take risks to improve their situation. Although they are well behind on the research curve, there is a very good chance that early adoption of nanotechnology, and thus the experience which could engender faster development, could well happen in Asia, Africa, or South America. This has a clear historical precedent in the way England outstripped France during the industrial revolution.

On the other side of the argument, we must note that rich West-erners still age and die, and they are likely to support any develop-ments that extend vigor and vitality. Nanotechnology could virtually eliminate physical handicaps, and ameliorate other kinds, another laudable and popular goal. We do seem determined to overeat, although we aren't so fond of diabetes and cardiovascular disease. Once nanomedicine begins to make inroads on any of these fronts, it's likely to pick up steam. However, nanomedicine is well behind genetic manipulation at this point, and probably will be for some decades—biotechnology simply has a big head start. So we may wind up with biotech, the more dangerous of the two technologies,[14] and not nan-otech, which could have helped with detection, shielding, and allevi-ation of biotech threats.

CHAPTER 11

TRANSPORTATION
It's a Very Small World, Indeed

I will build a motorcar for the great multitudes, constructed of the best materials, by the best men to be hired, after the simplest designs that modern engineering can devise. No man making a good salary will be unable to own one and enjoy, with his family, the blessing of hours of pleasure in God's great open spaces.[1]

—Henry Ford

B ack in the 1990s, Interstate 78 was finally completed across north central New Jersey. It had been held up for decades in the courts, and by the time it was finished, it cost over $30 million a mile, a high-water mark at the time for a nonurban road. In a twenty-five-mile-wide stripe centered on the highway, land values doubled virtually overnight.

Why? The new highway made it possible to get to the New York City area in an hour, when before it had been closer to two. The amenities of the city, ranging from Broadway plays to the major hub airports, were usable while living in a nearly rural setting where deer wander through your backyard and there's a buffalo farm (really!) next door.

In some parts of the country, like Alaska, private airplanes are more common than in suburban New Jersey. Planes have a couple of disadvantages compared to cars, though. They're expensive. They're less intuitive to operate and require a lot more training to use safely. They're more susceptible to bad weather. And they require a lot of space for runways and clearance from other airplanes.

Still, we fly routinely on airliners that go, on average, ten times as fast as cars. They're safer than cars, and cheaper for cross-country travel.

But traveling in airliners is a major pain. Consider a typical trip in the middle distance, say, five hundred miles. First, you don't get to pick when the flight will be; it may match your schedule, but in most cases you have to modify or plan your schedule to match the airline's. Then you drive to the airport and park, which takes an hour. Then you check in and wait, because you've been advised to be at least an hour early for the flight. You can't even carry a pocketknife. Everything you take has to be packed for each trip and will be inspected by officious strangers. I've had things like laser pointers disappear from my luggage in this process; I don't know if this was official overzealousness or private pilferage. Once the plane leaves the ground, everything is smooth until landing an hour or so later. Then it takes another hour, or more, to get your luggage, rent a car, and drive to your actual destination. And that's if everything goes smoothly, your luggage isn't lost, the flight isn't delayed, and so forth.

With a car, you just toss your stuff in the back and go. You can keep things in your car all the time rather than having to pack them for each trip, if you like. No poking, prodding, or searching—and no chance of being hijacked, either.[2] No lost luggage. You can smoke if you want, play loud music if you want, or have complete quiet if you want. Stretch out, rather than being packed like a sardine. Stop when you like and eat in a nice restaurant.

Since the 1950s, people have been tantalized with the idea of private aircars, which held out the promise of combining the best parts of both these modes of travel. If you could fly directly from your driveway, you could not only have the privacy and scheduling advantages of a car, but reduce travel time from four hours to one by avoiding all the inefficiencies of the airport. Even if your aircar is only half as fast as an airliner, you'll save a couple of hours.

Could nanotechnology build an aircar? There's no doubt whatsoever that it could, because current technology could build an aircar. The salient questions are whether it can reduce the cost, increase the safety, and make it quiet enough to use in residential areas without the neighbors getting up in arms.

Current aircars, such as the Moller,[3] are listed in the neighborhood of a million dollars each. Mass production could bring this down considerably, but it will take something like autogenous molecular manufacturing to make them really affordable. Given the strength-to-weight ratios of nanotech materials and the power-to-weight ratios of molecular engines, a powerful, capacious aircar need only weigh a few hundred pounds. This means a few hundred dollars in raw materials costs. In the long run, once development costs are amortized, an aircar could be significantly less expensive than ground cars are today.

Flying a light airplane can be a tricky business. The most common single cause of major accidents is euphemistically referred to by the FAA as "controlled flight into terrain," that is, diving into the ground when you thought you were flying level. It's perfectly possible to be flying upside down and not realize it, or to pour a cup of coffee while the plane is doing a complete 360-degree roll. Safe flying takes a lot of training and practice.

Flying a VTOL, a craft that does vertical takeoffs and landings, is even harder. Yet we want our aircar to be a VTOL. The reason is that a car-sized VTOL would be compatible with existing driveways, but a "normal" rolling takeoff plane would require you to build a runway. (Don't even think of a separate airport away from your house: it reintroduces all the airport congestion problems and requires you to have a separate ground car, not to mention the problem of getting a ground car at your destination.)

Thus the aircar needs to have an autopilot capable of doing all flight operations. These not only exist, but have been routinely in use on commercial airliners for more than a decade. An autopilot has not only the right reflexes for the conditions of flight, counterintuitive to humans, but is directly connected to an array of sensors that tell it a lot more about what's going on. For example, it could sense directly all the patterns of pressure and airflow around the craft. Nanosensors and nanocomputers for such an autopilot are simple for nanotechnology, whereas their current-day equivalents are expensive and high tech.

The dual constraints of vertical takeoff and automobile size make it quite difficult to design a machine that's quiet. Helicopters are obnoxiously noisy, and directed-thrust jets, like the Harrier, are ear-splitting. In simple terms, to go up you have to throw a column of air down. You can throw a thin column fast, like the Harrier, or a wide column more slowly, like a helicopter. The faster, the noisier.

The air thrown down by a helicopter, by itself, is like a strong wind, and you get a rushing sound but nothing overly loud or obnox-

Figure 17. An aircar taking off. Unfurling fancloth sails from extensible spars, a nanotech aircar would have a much quieter takeoff than any current VTOL aircraft.

ious. What makes the noise in a helicopter is the blades. They produce the concentrated whump-whump-whump you hear because they give the air a hard kick each time they come around, rather than moving it smoothly and evenly. The noise a helicopter makes, by the way, annoys the average person even when it's some 20 decibels quieter than the threshold of annoyance for smoother, more continuous sounds (like traffic). It's something like rock music in that respect (for some people!).

The quietest way to take off, however, is to jump. It's well within the capability of nanoengineering to design a leg that will extend to fifty feet in length and fold up into a pad less than an inch thick. This gives us two advantages. First, the aircar is already five stories up when the air moving thrust has to take hold, so the noise is farther away. Second, it's already moving at a good clip, so less acceleration is needed. It also turns out that moving faster makes aerodynamic thrusters (propellors, ducted fans, or jets) more efficient; you're throwing down a longer column of air and thus don't have to push it as hard.

A Harrier or helicopter landing is almost as noisy as a takeoff. There is a way to land vertically that is virtually silent, though: with a parachute. Taking that as an inspiration, we can design the next stage of our vertical flight. Unfurl a sheet of thin material over the area swept out by a helicopter's rotor. But rather than cloth, this is a mesh with holes about like window screen. In each hole is a tiny fan blade. Now you can move a large column of air without any whump-whumps. Instead of hanging by strings like a parachute, it'll probably be more controllable to deploy the fancloth from spars like a sailing ship. It could make for a fairly romantic vision: sailboats in the sky.

Once you get up to a decent altitude, a few thousand feet, you need to retract all the legs and sails and move fast. The first thing you do is extend wings. It's perfectly possible to stay up and move forward on thrusters such as ducted fans, but since you want to be moving fast anyway, wings make things more efficient. Typical airplane designs today have a lift-to-drag ratio of about ten. That means that the engines have to produce only one pound of thrust for each ten pounds of weight; the wings do the rest.

Virtually any current thruster style will work, in particular ducted fans; but there may be a way to do better. An object in motion tends to remain in motion; if your aircar were in outer space, it wouldn't need any thrust to keep going. In the atmosphere, however, there is drag. Drag is usually broken down into three categories, which correspond to the modes in which the energy is dissipated into the air around the plane. First is induced drag, which is associated with the lift. This results in the downwash of air behind the plane. It could be lowered by having longer, thinner wings like those of a glider. They, however, would increase the next kind of drag, skin drag. This is essentially friction of the plane's surface with the air moving across it, which heats the air. Third is form drag, which results in turbulence in the air. It can be reduced by using a streamlined shape.[4]

Rigid metal parts, as in current airplanes, can approximate streamlined shapes, but the streamlines change with every change in speed, attitude, wind, turbulence, and so on. Soft, compliant wings, like an owl's, can match the streamlines much more closely, so an owl can fly with virtually no turbulence and be totally silent to human ears. Nanoengineered wings will be able to adapt to the airflow closely and reduce drag and noise thereby.

Skin drag happens because surfaces aren't really smooth. When an

Figure 18. An aircar cruising. At altitude, legs, spars, and sails would retract. The configuration shown is optimized for negative drag, with long, forward-swept wings.

air molecule hits a moving surface, the surface tends to bump it in the direction the surface is moving, transferring momentum from the craft to the air. Atomically smooth surfaces could help, but even they are bumpy and atoms are sticky. Some momentum transfer will remain.

Suppose, however, you could put a paddlewheel or propellor on the surface and give the bouncing molecule a bump in the other direction? Then you'd have negative drag.[5] Then you could afford to put out long, gliderlike wings, because the drag isn't pulling you back, it's pushing you forward. So you can decrease the thrust necessary by a factor of five, since top-notch gliders can get lift-to-drag ratios of fifty. You've cut way down on induced drag, and get your thrust from negative skin drag.

If all this could be made to work, the power you need to fly would be 2% of your weight times your speed. Let's say 5%, to allow for inefficiencies in the negative drag, the power conversion, and so on. Even so, if you and your luggage and your plane together are half a ton, your 500-mile trip at 500 mph will cost you less than two gallons of gasoline (or less than five pounds of hydrogen).

Could it be made to work? Everything except the negative drag seems straightforward with the capabilities of nanoengineered materials and machinery. Shape-changing is too heavy and expensive for all but a few military aircraft today, but it is no problem for molecular technology; note that your body moves, and birds fly, by changing shape.

For shape changes such as extending legs, folding and unfolding wings, and extruding spars for sails, we can build structural members from very thin sheets of diamond. The layers are dadoed to each other so they won't pull apart and can only slide one way. These would be pulled across each other by millions of microscopic motors. A leg, for example, would simply be a set of telescoping cylinders like a radio antenna. It would just happen to be six thousand cylinders, each 5 microns thick, an inch long, and ranging from 6 to 8 inches in diameter. When retracted, it would form a doughnut an inch thick; when extended, a tube 50 feet long. Inflated to ordinary tire pressures, the pressure alone would lift over half a ton.

The brothers Wright steered their original Flyer by a process they called wing-warping. Alerons, flaps, slats, and so forth came along later. Wing-warping is a bit cleaner, less draggy, and a lot more agile: it's how birds' wings work. Just try taking off in your own length or turning in your own wingspan in a current-day airplane!

The aircar on the ground might be egg-shaped to cigar-shaped. The longer and skinnier, the more efficient the aerodynamics. For shorter ones, the back could extend in flight to produce a more streamlined shape. Wings, sails, and tail surface would be retracted on the ground, out of harm's way, and not blocking foot traffic near the machine.

If negative drag works, control surfaces could be even more minimal. Controlling the force (and possibly direction) of the drag could give you a more direct handle on the airflow patterns around the craft. To increase lift on a wing, make the drag on top more negative and that on bottom more positive. To turn left, use positive drag on the left wing and negative on the right. If it worked really well, you should be able to fly a simple lenticular shape ("flying saucer") whose surface had programmable drag.

The reason that negative drag isn't as straightforward as, say, shape changing is that it involves using nanomachines directly in an uncontrolled external environment. Tiny dust particles could smash your tiny wheels or propellors. What would be unnoticeable levels of moisture could immerse them in water. It might not be possible to find a design that overcomes all the difficulties and still gets the efficiency advantage we'd like. At this point, my analysis indicates a better than even chance it could be made to work, but it's not as definite as many of the other designs I've referred to here. If not, we'd just fall back on ducted fans, like airliners use today. Not nearly as elegant, but we know they work.

What about safety? The main causes of problems light aircraft can have are pilot error, weather, turbulence caused by large aircraft, and engine failure. Pilot error is minimized with the highly integrated autopilot and sensor system. Since it can communicate directly with other aircraft and radar networks, there's less chance of unexpected encounters. Since you're not operating near big airports, you can more easily avoid large planes. Inclement flying conditions are much more easily handled when the craft has a factor of ten or one hundred of power to spare compared with current models. Current-day aircraft are tossed around like leaves by high and turbulent winds, but nanotechnology will make it possible to put a much higher peak power capacity in the plane without making it too heavy or cost too much. That will improve safety. And finally, nanotech gives you the option of using millions of tiny motors instead of one or a few big ones. Any given engine failure wouldn't even be noticed.

Won't the air traffic control system be overburdened? Yes, it would, because it's an antiquated, bureaucratic mess. If you had to ask a "ground traffic controller" for permission at every turn when you drove, driving would be a similar mess. Instead we have autonomous control, with automated arbitration like traffic lights, and it works reasonably well. Watch a flock of birds land in a field and take off again. They are flying within a wingspan of each other at speeds equal to city automobile traffic, with nervous systems considerably less complex than yours. They don't bump or crash. Watch a cloud of flies around a garbage can. They perform the same feat with nervous systems your PC can match. No central control in either case. Purely distributed, autonomous navigation based on local information. If, with all our fancy technology, we can't do as well as flies, I'll eat my hat.

What about crowding, congestion? Over cities, certainly; so what's new? Outside them, we can estimate congestion by noting that there are 200 million cars in the United States,[6] and 4 million miles of roadway.[7] If we spaced the cars evenly along the roads, they'd be about 35 yards apart. Since there are about 50 million cubic miles of flyable airspace over the country, the same number of aircars evenly spaced would be over half a mile apart. That's if they were all up there at the same time.

Each American car is driven, on average, 10,000 miles a year.[8] Leaving out cities again, let's assume this is at an average 50 mph and takes 200 hours, less than 2.5% of the time. If we assume the aircars are traveling ten times as far on average, they're still taking only the same amount of time, so the average number in the air is about 4 million and they're over 4 kilometers apart. In other words, if you flew from New York to San Francisco completely obliviously, not watching for traffic at all, there would be one chance in 100 million you'd hit

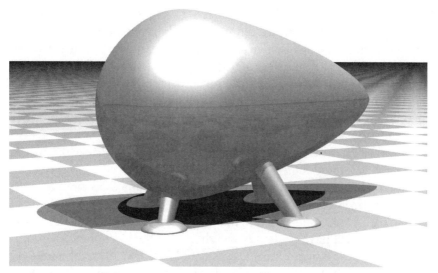

Figure 19. Home again. Back on the ground, the aircar would fold up all wings and sails and shorten from the aerodynamically efficient long shape to a garage-friendly short one. Shape-changing nanotech materials would make doors, windows, and so forth virtually undetectable when closed.

another car. Add radar, transponders, and an always-alert autopilot, and it's something I wouldn't worry about.

Over densely populated areas, you (or your autopilot) would probably have to file a detailed flight plan with some central authority, essentially taking out a reservation on each successive spot in the airspace you intend to occupy. This could be completely automatic, indeed completely decentralized, as traffic on the Internet is now.

Aircars, like most other powered machines operating in Earth's atmosphere, will probably use hydrogen for fuel. Hydrogen, as we have seen, is considerably lighter for the energy it gives you than hydrocarbon fuels like gasoline. That's useful in a flying machine.[9]

How fast will it fly? From 1920 to 1970, the speeds (and altitudes) of commercial airliners increased along a smooth exponential curve. Then they stopped. There are two reasons for this. First is that bigger aircraft can fly faster and more efficiently, but with the 747, diminishing returns set in as it gets more inefficient on the other side of the economic equation to make more and more people go to the same place at the same time.

The other reason is the sound barrier. The technology to fly faster than sound is half a century old now, but flying just over the speed of sound is three times as expensive in energy as flying just under it. So modern airliners fly just under the speed of sound and right at the edge of the stratosphere, six miles or more up.

It's quite possible to have a supersonic private plane. Several people do today, but as you can imagine, they're hellishly expensive. But the flight regime of the current-day airliner is a high point in efficiency. We can reasonably expect the design of the everyday flying car to follow the same logic.

At airliner speeds, Punxsutawney, Pennsylvania, is as close to New York in time as central New Jersey is now. And closer to Philadelphia, Pittsburgh, Cincinnati, Cleveland, and Toronto. Perhaps Phil the groundhog will have some new neighbors. If the I-78 phenomenon is repeated, and it's by no means an isolated case, the value of land over much of North America outside of existing metropolitan areas could rise considerably. That's the sign of a huge social benefit.

There remain reasons to want to go faster than airliner speeds. Suppose you want to visit Australia. From New York, it's twenty hours in the air and a bit more with stopovers. The Pacific leg is at the edge of the range of modern jetliners; I once was on a flight that had to put down in Fiji, unscheduled, to refuel because we had fought headwinds on the way.

The space shuttle, of course (or even John Glenn's *Mercury* capsule), travels that far in under an hour. The frustrating part of it is that it takes no more energy to get into orbit than it does to drag a plane through the atmosphere halfway around the world.

A passenger rocket in the nanotech era might look something like a modern-day airliner. Like any new technology, it would be commercial and concentrated at first, just as railroads preceded private cars. On the inside, though, would only be a first-class cabin; the rest would be fuel and oxygen tanks. The rocket would take off on a runway (we've got all these airports already anyway) and climb just like a current-day jet on ducted fans. Instead of leveling off, it would keep climbing. At takeoff, the rocket would have only hydrogen fuel aboard. As it flew, it would suck in huge quantities of air and liquify the oxygen for use in the rockets.

Nanotech could make rockets quite a bit safer and cheaper than they are today, but not too much more efficient; they are reasonably near the physical limits today. There are several possible ways to get the rocket effect besides the standard combustion chamber and nozzle arrangement. One way might be to burn the fuel to produce power and use that to accelerate the reaction mass electrically. Spaceborne ion thrusters work that way today, but don't provide enough thrust in their current form. It might be possible, burning hydrogen in oxygen, to accelerate the resulting water molecules without ionizing them, since they have enough of a polar electric character to grab electrically.

Most of the nanotech applications I talk about won't push the technology to its limits, so it's possible to give safe predictions by using designs we can currently simulate and analyze. Rockets do push the limits, as do any of the other possible schemes for reactive thrust. So we can't say with great assurance how they will work in detail— there are a great many ingenious innovations to be made. However, we do know how to make rockets today, so we know they're possible. We might make thousands of tiny rocket engines and be safe against any one failing. Alternatively, scaling laws favor larger ones. The best engineering compromise remains to be seen.

Your rocketliner accelerates up on fans for two or three minutes, loading its oxygen tanks. Then it cuts in the rocket engines and retracts its wings to a highly swept, sharp-edged shape; by this time it's at least ten miles up, and the noise, rockets and sonic boom, won't be nearly as noisome to people on the ground. You accelerate at less than two Gs for about ten minutes, coast for half an hour, brake in the air another ten minutes, and fly down to the airport.

How much would your ticket cost? One way to estimate is that current-day airlines operate at about six times fuel costs.[10] For rockets today, the technology costs much more proportionately, but nanotechnology could bring that down into line with current jet technology. Assuming your share of the weight is a quarter ton, you need the same weight in hydrogen, at a current cost of about $190 in gas form. There would be traffic at $1,200 a ticket but not high volume. If fuel costs got down to the level of coal as delivered to utilities, your ticket would cost $200 instead, and there would be plenty of takers.

FREIGHT

The cost of coal is illustrative of a more mundane aspect of transportation. Coal at the mine mouth in Wyoming costs $5 per ton. Delivered to utility-generating plants by the traincar, it's about $30 per ton. Delivered to your house in Maine by the single ton, $135.

Railroads are an old, established, highly optimized means of transportation. Nanotechnology could squeeze some more optimization out of the current hardware, by such means as self-maintaining track, faster trains, and smoother rides.

In the long run, though, surface transportation can only get more expensive, since it requires land. Whether train or truck, surface vehicles put the noise and danger right where people want to live. The same technologies used in the private aircar should be able to make airtrucks for payloads of up to ten tons or so, which wouldn't be too much noisier than current-day trucks. Airtrucks should be at least as energy efficient as ground trucks for long distances as well. Trucks use a large proportion of their energy stirring up the air, and a somewhat smaller proportion breaking up the pavement. Carefully tuned aerodynamic shapes flying over hills, valleys, and stop signs would lower energy costs.

A large part of the need for bulk transportation could simply be obsolete with the advent of general-purpose synthesizers, or even special-purpose ones in local stores. There has been a historical trend in this direction. When I was young, you took a roll of film and mailed it off to Rochester for processing, getting your prints back a week later in the mail. In the 1970s and 1980s, the technology for one-hour photo shops in malls and main streets was available. Now, of course, you take a picture with a digital camera and print it out if you want, or simply e-mail it without its ever touching paper.

Being able to review and save pictures in digital form has cut down on the amount of photo paper necessary, but some paper still has to be manufactured and shipped in bulk. Indeed, the computer and office printer, as well as the copier, have dramatically increased the amount of paper we have to contend with. What's missing, that nanotechnology could provide, is local recycling. Once the waste stream can be turned around locally, by home recyclers, long-distance transport of bulk materials (not to mention waste!) will begin to decline.

BEAM ME UP, SCOTTY

What about matter transmission? To be more specific, using molecular dissasemblers to take an object apart, noting the type, position, and bonding of each atom, and transmit the information to somewhere else, whereupon an molecular assembler puts together an exact copy?

Don't hold your breath. Let's suppose the high-bandwidth link to your house is a cable with a core that consists of one million optic fibers, and that the transducers are able to send data at optical frequencies, 100,000 times faster than today's gigabit fiber. Your total data rate is 100 exabits per second (i.e., 100 billion gigabits). That's a reasonable expectation for a nanotech data link of the later twenty-first century. It could transmit the entire contents of today's World Wide Web in a tenth of a millisecond.

The human body contains something like seven billion billion billion atoms, and to describe each one's type, position, and bonding you need something like 100 bits, even with compression. In other words, the time it would take to transmit a complete atomic-level description of yourself along your nanotech data link would be about 222 years. You might as well walk.

You could save a lot of data by transmitting a close, but not exact copy. Your body contains a lot of water; no need to transmit the position of each molecule, just say "water in here" and some notations about how salty it is, and so forth. Furthermore, many of the complex molecules are the same, or should be—proteins and DNA come as many copies of many fewer molecule types, and variations are injuries you'd just as well correct. It seems likely that if you allowed a little fixing-up, and didn't mind as much variation in atom positions as would happen in a tenth of a second or so, you could probably transmit the information required in a reasonable amount of time.[11]

There's one caveat about transmitting yourself this way, though: it's just as possible to make two copies, or even a hundred, at the far end as one. Heaven only knows what cans of worms we could open! But personally, I'm not sure I'd want to transmit a fairly good copy in the first place. Let's just say that teleportation is an iffy proposition: not obviously impossible, but not clearly practical, either.

On the other hand, transmitting the descriptions of manufactured objects will be routine. These will be hierarchically structured, built of many copies of identical parts, and much, much simpler than raw, natural objects. Transmitting full-sensory telepresence data will be easy also; you can get close to that with a gigabit link today, and 10 gigabits is almost certainly enough for complete fidelity at human sensory resolution.

Physical personal travel for business might then decline, because high-fidelity teleconferencing could be just as acceptable as real meetings, and a lot faster. Personal tourism might well increase, though. Bulk freight might decrease but personal items—handicrafts, mementos, souvenirs, and other objects whose value lies in their identity—might increase. The mailbot will still bring a pound of unwanted paper trash each day but you'll just toss it into the recycler.

CHAPTER 12

SPACE
Where a Much Larger World Awaits

I t was 1969, the year of *Apollo 11* and Woodstock. The tumultuous 1960s had just seen a countercultural revolution and a widespread disaffection with science and technology, especially on university campuses where government-sponsored research was associated with the Vietnam War. At Princeton, physics professor Gerard O'Neill, inventor of the particle storage ring, had drawn the rotating duty of the freshman physics course and was looking for some way to make it socially relevant.

That was also the year that the San Francisco city council adopted the Earth Day holiday, which was then celebrated on the vernal equinox in 1970. In other words, environmentalism was becoming relevant. So O'Neill put all this together and challenged his students: "Is a planetary surface the right place for an expanding technological civilization?" And if not, the unstated subtext implied, what was?

O'Neill and a small cadre of interested students got to work, and over the course of the next few years, an interesting and somewhat unexpected answer began to emerge. In science fiction, the consensus view that had developed into an orthodoxy over the first part of the century had assumed that we would eventually settle onto the surfaces of other planets. But O'Neill's group started out, perhaps by the accident of the wording of the challenge, with a different perspective. They found an alternative: build living places in space, from scratch. And the more they looked at it, the more the numbers seemed to work out.

As the 1960s turned into the mid-1970s, the idea had caught on and the designs took on significant detail. The Club of Rome's *Limits to Growth*[1] had come out; the energy crisis was in full cry; environmental concern was increasing; the Earth's population seemed ready to burst the seams of this small, overburdened planet.

"O'Neill colonies," as they came to be called, would be miniature, inside-out worlds. They would rotate to provide gravity, and since the ground would be on the outside, it would serve as a radiation shield in the place of Earth's miles-thick atmosphere. Plans and artists' conceptions called for lush garden communities; lots of plants would help recycle the air. Farming and industry would flourish nearby in separate structures.

In space you can set out mirrors thinner than household aluminum foil, and they'll just stay where you put them, with no wind or weather and the completely predictable microgravity of orbit.[2] Such mirrors collect a completely clean, uninterrupted, totally reliable solar power. For example, a burner on your stove in the space habitat could be powered by mirroring in sunlight from just a couple of square yards outside. No coal mining, no atmospheric effluents, no nuclear power plants, no cross-country power grid; just a collector the size of a golf umbrella.

The material to build the habitats would come mostly from the Moon, whence it was an order of magnitude easier to get into the orbits of interest than material from the Earth. Lunar soil was being analyzed by the follow-up Apollo missions and found to contain almost all the necessary elements—only hydrogen would have to come from Earth. High-energy smelting and other industrial processes, too costly to use on Earth, would be feasible with the bountiful, essentially free, solar energy.

As the ideas gained popularity, an increasing number not only of enthusiasts but of serious engineers at MIT, Stanford, NASA, and various aerospace companies began doing analyses and contributing ideas. (Among them, by the way, was a young K. Eric Drexler, who wound up doing his master's thesis at MIT on vacuum vapor-deposition space manufacturing.) Not only were space colonies feasible in the known technology of the 1970s, but they could pay their own way: solar power could be harvested and sent to Earth, replacing fossil fuels and obviating a host of environmental problems.

Most amazing of all, space colonies represented a flat-out, categorical solution to the problem of overpopulation. The trick is that space colonies, complete with people, can be self-reproducing. That is, the people of one colony build another one, which is populated by people from Earth. Each of these builds another, and so forth. The numbers showed that by building in the asteroid belt, there was enough easily accessible material of the right kind to build colonies with twenty-thousand times the habitable area of the Earth. (After which you can begin looking at moons, planets, comets, etc.)

In the thirty years since these ideas were developed, the Earth's population has risen from 4 billion to 6 billion. That's an annual increase of 1.4%. Increases in poorer nations tend to be higher, in richer ones lower. The United States has averaged about 1% recently.[3] If this rate could be made to apply worldwide, by raising the standard of living, for example, keeping the population static would require moving 60 million people off the planet annually. For comparison, the three New York area airports handled 92.6 million paying passengers in 2003.[4] In other words, three large spaceports could handle the Earth's population growth at US growth rates; at third world growth rates, you'd need six spaceports.

The bottom line is that space colonies represented solutions to quite a few of the world's problems as seen from the 1970s: population (and hunger, since the colonies are self-sufficient in food); the energy crisis; pollution; and to some extent war, if fought for lebensraum. At its height in about 1977, it could have been called a movement, backed by the ten-thousand-member L5 Society and numerous public figures. What's more, according to a wide sampling of leading scientists and engineers, it was within the technological capabilities of the day.

So what happened?

> As space visionary Arthur C. Clarke observed more than a quarter of a century ago, operating expendable space transportation vehicles is equivalent to building the ocean liner H. M. S. Queen Elizabeth II, sailing it once across the ocean, and scuttling it upon arrival at its first port of call.
>
> —Harry Stine, *Halfway to Anywhere*

Space colonization foundered on the rocks of several realities. Largest and most jagged was the issue of the cost getting to orbit. In the 1970s, the space shuttle was on the drawing boards, and NASA confidently predicted that it would lower the cost of access to space. The cost in the 1960s, using the Saturn V moon rocket booster, which was thrown away as part of the launch process, worked out to $3,800 per pound of payload. Surely, with a reusable space truck we could do better. Then came the inevitable bureaucratic reality. By NASA's own extremely optimistic figures in the 1980s (before any explosions), the cost of shuttling to orbit was $6,000 per pound. More realistic accountings by independent analysts put the cost as $20,000 to $35,000 per pound. Shuttle proponents had promised O'Neill that it would reduce costs to $430 per pound. Woops.[5]

Next was the fact that a lot of the eco-angst of the 1970s turned out to be way overblown. The energy crisis was, in the medium term anyway, a product of OPEC price manipulations and not a sign of impending doom. Higher oil prices encouraged the development of new sources of supply, as higher prices will. The cartel collapsed in the face of competition, as cartels will. Although nuclear power became a political whipping boy in the United States, it was developed into a major energy source in places such as France. The "green revolution" and continuing improvements in farming methods and crop yields kept pace with the population and prevented any famine not caused by intentional political intervention. The predictions of *Limits to Growth* and its ilk simply didn't come true: we were supposed to have run out of oil and "many key minerals" as long ago as 1985.[6]

And finally, even as envisioned by the would-be space colonists, the project would have required a national effort of Apollo Project magnitude for twenty years. The political will simply wasn't there.

Are we, then, stuck on Earth? (Not being able to move away counts as stuck.) The answer is no, we're not stuck. What was technologically feasible thirty years ago is feasible now, and will be even more so in the future. Although we're not about to starve in the dark here on Earth, land prices do keep rising. And people do seem to keep worrying about fossil fuels. So at some point it seems quite reasonable to predict that the trends will meet to make space-based land a better buy than Earth-based.

Another point is made by a quip that was heard in the early days of aviation: "Airplanes breed like rabbits, and dirigibles breed like elephants." The bigger a project is, the harder it is to get off the ground, even with the same rate of return. So the point where space colonization really becomes a force is most likely to be when a moderate-sized group, say 250 people, can pool their resources and build themselves a new home, not unlike the groups involved in crossing the Atlantic in tiny sailing ships from Europe or crossing the West in Conestoga wagons from the East.

What kind of a dent could nanotechnology make in the problem, and how soon? Let's imagine it's about fifty years from now, that is, the amount of time that passed between Kitty Hawk and the 707. There are plenty of things nanotechnology hasn't done yet, but substantial progress has been made on several fronts. In particular, let's assume most of the progress is not specifically aimed at space colonization at all, but at meeting very ordinary terrestrial needs.

Foremost is nanomedicine. Several kinds of diseases involve intracellular damage at the molecular level, from things like free radicals and radiation. Osteoporosis involves calcium loss from bones, and it is likely treatable by some relatively minor chemical and electrical props at the right places in the body. People will very likely want to maintain fit and healthy bodies without having to exercise every day. Nanomedicine will very likely have addressed these issues.

This will make the design of a space habitat considerably easier (not to mention cheaper). The single most massive part of an O'Neill-style colony is the radiation shield, six feet of lunar soil surrounding the living space. With a higher radiation tolerance given by cell repair machines, a much less expensive shield provided by a magnetic field and a small shelter for major solar flares should be more than sufficient.[7]

The other major design constraint was due to having to rotate the structure to maintain artificial gravity. Nanomedicine addresses this as well. It seems quite likely that the human body could be maintained in fit shape in microgravity with the same kinds of interventions that will already be popular to keep fit without exercising.

Avoiding both these constraints gets us out of the economic land of elephants and back among the rabbits. Much smaller habitats become feasible. Elaborate systems of mirrors to channel sunlight past radiation shields disappear. Heavy construction necessary because of the artificial gravity disappears. The necessity of building large circular structures that are rotationally symmetric disappears.

Farming and food production in space were among the easier of the problems that the O'Neill studies looked at, partly for the same reasons: plants are typically radiation resistant and not bothered by weightlessness. Even without nanotech help, space agriculture seems feasible and economical. With nanotech help, such as direct synthesis of some nutrients and some foodstuffs, assisted breakdown of waste products, and so forth, it's a very moderate item in the budget and on the chore schedule of the colonists.

In fact, nanotechnology improves the economics of the space habitat so much that it's almost viable to bring the building material up from Earth. But mass from the Moon will still probably be only 1% to 5% the cost per pound. So you'd most likely bring up the nanomachines (and hydrogen) from Earth, and buy raw material from the Moon.

Compared to the ten-thousand-person designs of the 1970s, the technology of 2050 will make possible single-family dwellings in space, completely self-sufficient, and easily within the economic reach of an average American family of the period, if current economic trends hold true. This might well be less than the price of a comparable house on the ground in many parts of the United States or Europe by that time. Of course by that time you'll have other options as well, from mountains to high latitudes to seaborne to submarine living areas. But even those will start to look crowded in time.

Besides the things that nanotechnology allows us to leave out of such a house, what does it allow us to put in? Windows, for one thing. Large panes of transparent material with a strength and toughness suitable for structural applications. A decided advantage will be the ability to join such panes (and other sections) with no compromise in structural integrity. Another will be that all the hull material, not just windows, will be smart. It will need to be able to change its optical properties, not so much for lighting control as to regulate temperature. It will need to check its own condition constantly, be able to do at least minor self-repair, and raise an alert when a major repair is needed.

Inside the house, you'll find the same kinds of things that nanotechnology will be providing for people on Earth—virtual reality walls or booths, robots, and the whole infrastructure of the information economy. You'll also find a much more thoroughgoing manufacturing capability than would be usual in a terrestrial home, just as you find more mechanical do-it-yourself capabilities on a farm than in a city apartment now. Apartment dwellers of today rarely need to weld a broken tie bar on a tractor; those of tomorrow will be as unlikely to need to recycle a spacecraft.

The main improvement over current practice in space, though, is the spacesuit. This is something Drexler worked out in the 1980s and described in *Engines of Creation*. It starts out a lot like the skinsuit, but thicker. It should have the same "feels like nothing" interface to your skin, but instead of protecting you from a temperature difference, it protects you from a pressure difference. To this end it is thicker, maybe a millimeter or so on the fingers to an inch or two on the back, forming a storage area.[8]

The reason the suit must be thicker is that the atmospheric pressure it contains exerts quite a lot of force. Over the approximately two square meters of skin area on a human, the 100 kiloPascal pressure of air on Earth exerts a force of 200 kiloNewtons, that is, 20 tons. You don't normally feel this because it's distributed equally over your body in all directions, and your body is mostly water, incompressible for all practical purposes. But the spacesuit has to hold that much force without anything pressing back from the other side, so it will try to blow up like a balloon.

Current-day spacesuits are very difficult to move in. Take an inner tube and inflate it to 15 psi, as if you were going to use it for a pool toy. Now try folding it in half. Imagine doing that every time you moved your arm! When you fold the inner tube, you compress the air inside to a smaller volume; release it, and it pops back to its original shape and volume. To try to help, current-day spacesuits are used with as little pressure inside as the human body can tolerate, something like 20% of sea-level pressure (that low, it has to be pure oxygen). Even so, moving and working for an extended period is quite clumsy and very tiring.

Your body moves effortlessly in sea-level pressure here on Earth because when you move, your body changes shape but not size. If you expanded or contracted, you'd have to work against that twenty-ton weight. The solution for a spacesuit is the same: maintain a constant volume inside, and there's no force favoring one configuration over another. Because your body maintains a constant volume, the suit can do the same just by hugging your skin exactly. (Breathing is a separate issue, and will require a separate solution.) But to conform to the shape of your skin, the suit needs a lot of strength to keep from blowing up into a round balloon shape. What's more, it has to maintain this strength while changing shape as you move!

A spherical balloon of diamond with two square meters of surface area would need to be less than half a micron thick to hold sea-level pressure. The hardest shape to hold pressure with is flat. A flat diamond sheet holding sea-level pressure across a foot-wide gap would have to be almost a millimeter thick—two thousand times as thick as the balloon! A one- or two-centimeter thickness over wide flat areas of the body, like the front and back of the torso, gives us plenty of margin for safety, as well as room to put lots of machinery inside along with the load-bearing structure.

In fact, all the load-bearing structure is active. The suit is, in effect, a hollow robot that you just fit exactly inside of. It gets its commands from pressure sensors touching every square millimeter of your body, and responds so fast and precisely that you feel no resistance to your motions at all.

Besides pressure, a competent spacesuit must provide other aspects of the Earth's environment. The skin needs oxygen and it needs for sweat to evaporate. It needs protection from the strong ultraviolet (UV) light of the Sun, but it needs UV in moderation for vitamin D synthesis and as an input to the circadian rythm. The suit, if properly designed and programmed, can provide all of this. It can bathe you and sunbathe you, remove mites and bacteria physically instead of having to use chemicals, and even give you a massage.

For a "helmet," the simplest of many options is to have a flap of material in a bubble shape that attaches around the neck. It could inflate to a transparent sphere in use, and roll itself down into a collar when you go inside, like the hood of a windbreaker.

Like the walls of the house, the suit regulates internal temperature by controlling incoming and outgoing radiation (sunlight). And one of its major functions is self-monitoring and self-repair. It should be about as comfortable to wear indoors as current-day inert clothing, so it should make living and working in a space environment as convenient as in a temperate environment on Earth.[9]

GETTING THERE

The major hurdle to getting into space is economic. By its own figures, the government would have had to do the equivalent of giving each of the L5 Society's ten thousand members a million dollars a year for twenty years for the O'Neill colonies to become a reality.[10] Current NASA launch costs are too high for economical space coloniza-tion by a factor of about one hundred. It's like sitting at home on Sat-urday night because cars cost a million bucks. What could we do to alleviate this problem?

The first step is building space vehicles that aren't thrown away each time they're used. This can be done two ways: either build a single spaceship that starts, flies, and lands as one piece, the way your car does; or build one that stages, as current rockets do, but in such a way that each piece flies safely back to be reused.

With the space shuttle, NASA tried in some sense to do both. The shuttle orbiter is a technological tour-de-force of the kind needed to make a single stage to orbit (SSTO) vehicle, burning hydrogen instead of the easier-to-handle kerosene, having rocket engines of an unprecedented efficiency. The solid rocket boosters (SRBs) are jetti-soned in flight but parachute to the ocean and are recovered. Only the external fuel tank is thrown away ("only" though it is the biggest part). The problem is that they pushed the technological limits too much, and the shuttle needs a thorough and very expensive overhaul after every flight—and even then it occasionally blows up.[11]

Nanotechnology will probably make a true SSTO possible. But it's not necessary to reach that technological goal to begin with. Since 1990, Orbital Sciences Corporation has been routinely putting small (one-thousand-pound) payloads into orbit with their Pegasus air-launch system. This means dropping the rocket from a high-flying airplane just as Chuck Yeager's X-1 was in 1947, when he first broke the sound barrier.

To get into orbit, you have to do two basically different things. You must go up to a height that will put you out of the atmosphere: 60 miles is good, 120 is better. Then you need to accelerate to about 5 miles per second going sideways. Rockets typically do this by starting out straight up and curving over until they are both high enough and fast enough. The engineers who design the flight paths have to juggle a lot of trade-offs. The higher the acceleration, the less time the rocket has to waste fuel just holding itself up during the ascent;[12] but the harder it is on the occupants, and the rocket itself has to be built heavier to withstand the force. If you start fast, you add a lot of drag in the lower atmosphere, all of which has to be overcome with extra fuel.

An air launch finesses this to some extent. An airplane held up with wings can get more than twelve miles up (the U2 does, for example). At this point, you're already above more than 90% of the atmosphere and can launch sideways. You can put small wings on the spacecraft (as the Pegasus has) to let the air help hold you up while you're accelerating, so you can accelerate more slowly. That means smaller, lighter engines and a lighter frame, as well as more comfort for your passengers. Your altitude has aleady accounted for 10% to 20% of the height you need. Your rockets engines are more efficient because of the lower pressure. The bottom line is that only 85% of the starting weight of your rocket has to be fuel, as opposed to 95% for a ground-launched SSTO.

That makes a huge difference. If you and your family, with luggage, weigh half a ton, and your spaceship, including cabin, fuel tanks, rocket motors, wings, and all, weighs a ton empty, then loaded and fueled, you're ten tons. For comparison, the Boeing 737, a relatively small airliner, has a payload of fifteen tons. A drop aircraft is thus easily within current capabilities, much less nanotech ones.

How much is this going to cost you? With a nanotech factory, you can take hydrocarbons like natural gas and use the carbon to build things, while separating the hydrogen for use as fuel. You can also take carbon in a more concentrated form such as coal. Current commodity prices for natural gas are less than 50 cents per kilogram, and coal is about $30 per ton. Coal isn't pure carbon, so we might need two tons for the one-ton spaceship. We need 50 kilograms of hydrogen, requiring 200 kilograms of natural gas. With a markup factor of one hundred for the spaceship, it costs $6,000. Liquid hydrogen today costs something like $10 per kilogram, and liquid oxygen about 10 cents. We need 50 kilograms of the former and 800 of the latter, for a cost in today's prices of $580 per flight. I'd expect this to become lower since hydrogen is likely to become a widely used fuel over the next few decades, with competition driving the cost down.

The actual price of spaceships will consist almost entirely of amortized development costs, with raw materials counting for very little. Engineering costs for new rocket engines can be in the $10 million range today. (Note that by this point the liquid-fuel rocket engine will be more than one hundred years old—great works of innovative genius will not be necessary.) Assume engineering for the whole vehicle at $100 million: if you can sell a million ships, you need only add $100 apiece; if you only sell 100, you'll have to charge $1 million each. The bottom line: if enough people are interested, cost is not a problem.

LONGER-TERM OPTIONS

Once there is a reasonably high volume of traffic to space, costs can be reduced even further. One of the schemes you'll probably hear about is called the "space elevator," "skyhook," or even "beanstalk." You'll hear about it in conjunction with nanotechnology, even nanoscale technology as done by fullerene chemists, because it's a really nifty, compelling idea that is unfortunately not feasible with any material except perfect diamond fiber or fullerene nanotubes (bucky-tubes).

Here's the nifty, compelling idea. A satellite placed in an orbit of radius 42,164 kilometers (i.e., 22,000 miles above sea level) will orbit the Earth in exactly twenty-four hours, and thus can appear to hang stationary above a given spot on the equator. This fact is much used by communications satellites today; the orbit is called geosynchronous or Clarke orbit, and it's already somewhat crowded with satellites. (Because they stand still, so can your satellite dish.)

Now imagine that you are standing on a geosynchronous satellite, and you let out a twenty-two-thousand-mile-long string. Someone could tie a package to it, and you could pull it up, and bingo, you've put a package into orbit without using any rockets. In fact, if you put the string over a pulley and let one package down as the other comes up, you don't need to spend any net energy at all! Skyhook concepts typically are a lot more involved, with elevators, and counterweights on the satellite to balance the weight of the string, and so forth, but that's the basic idea. And what's more, in theory, it would work.

That's in theory. In practice, the thing is huge, if narrow. If it breaks, the cable comes whipping down into the atmosphere like a twenty-thousand-mile-long meteor, causing widespread consternation. Probability of breaking: 100%. Any satellite that is not in geosynchronous orbit is guaranteed to hit it, eventually, at speeds on the order of five miles per second. According to the US Space Command in 2000, 8,927 human-made objects were in orbit.[13] Even if we cleaned these up and prohibited any new satellites, problems remain. Because the skyhook has to be near the equator, it goes through the

Van Allen radiation belt. You're going up at elevator speeds, not rocket ones, which means you get to soak up quite a lot of radiation on your way up. If the elevator runs at 100 mph, it takes more than nine days to get to geosynchronous orbit.

Rockets breed like rabbits; skyhooks breed like elephants—and the two methods are mutually incompatible. By the time technology and economics make skyhooks possible, there'll be too heavy an investment in satellites to change course.

A somewhat more realistic idea can be had by looking at ships. Ships move well in water, but have a pretty hard time on land. Typically when you need to get on and off a ship, you don't beach it, but bring it up to a pier. A pier is an extension of land built so that a ship can operate in contact with it without having to be beached.

How do we build a pier for spaceships? The equivalent of the beach, for Earth, is the atmosphere. If we build a tower twelve miles high, it gives us the same advantages as the air launch. If it's long, as well as high, we can help the rocket by putting a mass driver on top. A mass driver is essentially a cannon that uses magnetism instead of an explosion, so the "cannonball" can be something relatively fragile, like human bodies.

How much can we help? In fact, if we make the tower 60 miles high, and 180 miles long, we don't need the rocket at all. You're above all but one-millionth of the atmosphere. Accelerating in the mass driver at 10 Gs, an acceleration healthy humans can handle, will put you into orbit. Sixty miles is high enough for orbital speeds in the

Launch Tower to Orbit

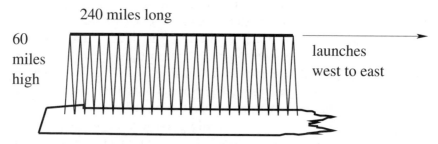

Figure 20. A space pier. A launch tower is shown standing on Pennsylvania for scale. On this scale the curvature of the Earth is not noticeable.

mass driver, but low enough not to bother satellites. A 100 mph elevator ride to the top of the tower takes only thirty-six minutes. The launch through the mass driver takes less than a minute and a half.

For freight, we don't have to stop at LEO (low earth orbit). Mass drivers delivering 30 Gs were built in the 1970s. Thirty Gs on a 180-mile track gives the ship enough oomph to get to Venus or Mars; the Moon and L5 require only 20 Gs. People would have to make a stop in orbit to be accelerated again on a much longer driver.

Whether the tower will be worth building depends on the volume of freight to space and the cost of electricity. Running the mass driver continuously (and assuming ten-ton spaceships) would put nearly four million tons of shipping into space each year, and require 4 gigawatts continuous power. Nanotech-based solar power satellites can probably be designed at about 10 megawatts per ton, so the tower needs to launch forty for its own use—that will take an hour. I can't imagine the tower itself costing less than $10 billion, so they would represent a trivial fraction of its cost as well. If the tower did cost $10 billion, and was used at full capacity, amortized cost would be under $100 per ton, which is comparable to current international seagoing freight rates. Note that over five million tons of freight is shipped annually on the Snake River alone; most major US waterways handle one hundred times as much.

CHAPTER 13

ROBOTS
For Whatever Work There Is to Be Done

Back in the 1940s, the young Isaac Asimov took a staple of science fiction, the anthropoid robot, and made it his own. His series of robot stories and novels defined for the rest of the century, but particularly for the futuristic 1950s, what a robot should look like and what it could do.[1]

Physically, a robot was a machine made of steel in roughly the shape of a man. It had machinelike properties: it was strong, didn't get tired, didn't feel pain, and didn't need to eat.

The real robots of today are typically made of aluminum, plastic, or composites, as well as steel; but the ones of tomorrow will be gemstone balloons or banks of fog.

Mentally, the robot had a positronic brain. This was just glitzy science fiction verbiage. Positrons had recently been discovered, so the obvious science fictional step past electronic brains was positronic ones. In reality, as a particle of antimatter, a positron would be a bull in a china shop in any actual machine, annihilating itself and the first electron it ran into and releasing hard gamma radiation.

Skipping lightly over that, then, Asimov patterned his first robot after an animal. It could understand a limited amount of language but couldn't talk. It wasn't too bright, but was trustworthy like a dog.

Later experience with real machines has shown us that talking is easy but understanding is difficult, something we have long suspected from our dealings with human beings.

Robots today are much more prevalent than people may realize.

Your car and your computer were likely partially made by robots. Industrial robots are hugely expensive machines that must operate in a carefully planned and controlled environment, because they have very limited senses and no common sense whatsoever. With nanotechnology, that will change drastically. Indeed, it's already starting to change, as the precursor technologies such as micromachines begin to have their effect.

Existing robots are often stupider than insects. As computers increase in power, however, robots will get smarter, be more able to operate in unstructured environments, and ultimately be able to do anything a human can. They will find increasing use, as costs come down, in production, in service industries, and as domestic servants.

The physical body of a useful robot is easily within the capability of current technology. A hobbyist can make one capable of performing a variety of household tasks for a few thousand dollars in parts. The sticking point is the controller—a computer that reads the sensors and tells the motors what to do—and the sensors. Robots are getting more common almost by the minute nowadays because the cost of computers, signal acquisition circuitry, video cameras, microphones, miniature gyros and accelerometers, pressure sensors, and the like are falling through the floor.

Once the sensors are all there—and today they cost more than the bones and muscles—there remains the problem of programming the controller to interpret them and drive the actuators so as to achieve useful, coordinated activity. This is not a completely solved problem, but you might be surprised at the progress that's been made in just the past decade.

The American Association for Artificial Intelligence holds a "robot challenge" among other robot competitions at its annual conference. These are for real, autonomous robots, not remote control drones. The challenge requires a robot to "start at the entrance to the conference center, find the registration desk, register for the conference, perform volunteer duties as required, and report at a prescribed time in a conference hall to give a talk." All by itself. At the 2003 meeting, two robots, one from Carnegie Mellon and one from Washington University, completed the challenge.[2]

With nanotechnology, robots could have a sensory nervous system as complex and sensitive as the human one. That means millions of sensors where current robots have to get by with less than one hundred. Nanocomputers will provide plenty of processing power to integrate and interpret what the sensors are sensing. Just as radical an advance will be available on the motor side. Instead of the ten or twenty motors that today's awkward robots must get by with, nanotech robots could contain hundreds, making them as supple and graceful as a person, or thousands, giving them the flexibility and fluidity of an octopus.

A balloon made of thin sheets of diamond, the size and shape of a human arm (with internal partitions to maintain the shape), would need fabric just a couple of microns thick to be as strong as the arm. Make it 5 microns to allow for foldable structures, motors, local controllers, communication and power networks, and so forth. Don't stop with the arm—make a full human-sized balloon. That's your robot. Uninflated, the fabric occupies 10 cubic centimeters (a ballpoint pen), and weighs less than an ounce.

The robot might need ballast to stay on the ground in a stiff breeze, but otherwise it's physically the size and strength of a human,

Figure 21. Stretchy stuff. Shape-changing of material can be made of sheets of diamond dadoed together, with motors embedded to make it slide the way you wanted. This would be used in everything from flying cars to robots.

probably a lot more flexible and graceful, and tireless as long its fuel supply lasts. If folding up was not an issue, you could fill the balloon with fuel—it would still weigh less than a human—and run for a very long time indeed.

In the old science fiction novels, a robot who wasn't being used would stand in a niche in the wall, like a statue. In reality, you'll hold out your hand, it will collapse to the size of a pen, and you'll put it in your pocket. That is, if it doesn't simply vanish.

UTILITY FOG

Nature wants stuff
To vie strange forms with fancy . . .
—William Shakespeare, *Antony and Cleopatra*

The physical properties of matter, how hard or strong something is, whether it is a solid or liquid, depend on how its molecules react to forces that are applied to it. If they resist any force that tries to move them relative to each other, it's a solid. If they allow very small forces to push them past each other but still try to maintain a fixed distance to their neighbors, it's a liquid. If they try to push apart no matter how widely spread they already are, it's a gas.

Soft solids, like foam rubber, resist distorting forces weakly but with a force that increases with distance stretched. Hard ones like diamond resist with a force that increases enormously with even the tiniest microscopic change in shape. Metals resist up to a point and then flow like fluids. Modeling clay is similar, but the transition point is a much lower force.

Modeling clay and foam rubber, of course, are not simple materials. Their properties derive in part from structure well above the molecular scale. The same is true of wood, paper, bone, muscle, concrete, and indeed most common materials besides plastic. Even metals have grain structure that causes their properties to vary. Heat treatment that changes the grain structure can double the strength of the metal.

If you can stand some graininess, you can often get a lot of flexibility in return. Look at your television screen: the picture is composed of dots that can change color. The dots don't move, but the picture can appear to as the pattern of color moves across the dots. If all the dots change color at once, one picture can vanish and be replaced by another instantly.

The dots on your TV screen have a size, and a speed of changing, that is designed to be small and fast enough to maintain an illusion of a moving picture to your eyes. It generally works, even though you can see the individual dots if you look closely and see a flicker out of the corner of your eye. If you swap the 300,000 dots of the TV for the 130 million of a high-quality laser-printed page, the illusion becomes almost perfect.

With nanotechnology, we can make robots the size of those 1,200-to-the-inch dots on the laser-printed page. These robots can change properties so as to simulate the range of properties of ordinary matter, resisting a push or moving in the direction pushed, simply by running a different control program.

Take a box full of such robots. You now have the solid equivalent of the TV screen, where each three-dimensional dot can take on properties that are not just colors but the physical characteristics of matter. Like pictures on the TV, tangible physical objects can appear, interact, and vanish with a broadcast command telling each robot which program to run. Unlike TV pictures, the virtual objects could pick up and carry real ones, have weight, be hard, soft, springy, or even liquid. They could do real work in the real world.

Fill your entire house with the robots, instead of just a box, and you'll be able to have furniture, appliances, robotic servants, and even your clothing appear and disappear on command.

The stuff consisting of a mass of such robots is called Utility Fog.[3] When I invented it circa 1990, it was the first "swarm robot" concept with "instant-on" virtual objects. Since other writers, notably Ray Kurzweil,[4] have opined that it should have a better name, I'll at least explain where "Utility Fog" came from. First, the stuff fills the air like fog, and you walk around in it. It would look a lot like fog as well. As for the "Utility" part, remember the *Batman* TV series from the 1960s? Whenever Batman needed some gadget, lo and behold, there it was in his "utility belt." There seemed to be no end of what the belt could produce—and it was always right at hand.

Living in Utility Fog might be a lot like being in a campy, tongue-in-cheek TV show, or even in a cartoon. Cartoon characters are constantly reaching up and grabbing guns, hammers, anvils, suits of

Figure 22. Utility Fog. A Foglet and a layer of them with arms linked. The substance would be formed of many thousands of such layers, each Foglet too small to see with the naked eye.

clothing, false beards, tables, chairs, signs bearing witty remarks, and the occasional steam locomotive out of thin air.

What about walking around in it, though? Wouldn't it be like trying to walk around in modeling clay? No, because the vast majority of the stuff would be running the "simulate air" program. Even though it is denser than air, about the same density as balsa wood, you wouldn't feel it because it's powered and would actively get out of your way. There are lots of options for the specifics of the human-Fog interface, but one that seems reasonable is that it would maintain a bubble around your head, like a spacesuit helmet.

"Foglets," the individual robots, on the inner surface of the bubble would be equipped with phase-controlled light emitters, so they could project a real-time hologram of what you would expect to see around you, instead of looking into a fogbank. Of course, the Fog would have a lot more information about what's out there than just its surface color—it's touching everything in sight—so it could give you false-color images revealing other properties or make objects appear to be translucent so you could see through them, or whatever you like.

Or it could display the appearance of an entirely different location and simulate the feel of any object there. Suppose I'm standing in an empty cubicle—empty, that is, except that it's filled with Utility Fog. There happens to be a high-speed data link from my cubicle to your living room. The Fog bubble around my head shows me a view as if I were sitting across from you, having afternoon tea. So far, no difference in principle from having a video camera and my watching a screen. But I decide to sit down, the Foglets around me getting out of the way to let me move, and the seat of my pants gets to the point where my eyes tell me your couch is.

The Foglets at that point aren't simulating air, but couch. They feel like a couch. They hold me up like a couch. The Foglets in front of me are simulating the coffee table, and I can put my teacup down on them a little too hard, hear a thump, and observe a splash. The teacup will work best if it's real and really there, and the tea itself *must* be real if I'm to drink it.

That's telepresence—half of it, anyway. What's happening in your real living room?

In the simplest case, a robot sits there and transmits what it feels to my cubicle, so that the Fog can produce the same sensations for me. This might be an ordinary robot, but it could also be a Fog robot—a mass of Fog the same size and shape as me. This would have the advantage that when I "left," the batch of Fog could be put to other uses. But the most interesting case would be if your living room were completely full of Fog, too.

Now, instead of having to walk into your living room in robotic form, I can suddenly appear there, as the Foglets in a certain area quit simulating air and start simulating me. I can do anything there that I could if I were there in person—Utility Fog is at least as strong as human flesh—pick up objects, open and close doors, drink tea.

But I don't need to drink actual tea in your living room, I have my real teacup in my cubicle. So the teacup in your living room that I seem to be holding can be made of Fog, too, and the tea as well—since I'm not there to really drink it.

Let's suppose that the chair you're sitting on is really there in your living room, but the coffee table isn't. It's a virtual object formed by the Fog. Here we sit, apparently together, apparently in the same place, but we're actually in two different places. Some of the objects we're using are actually with you, and are being simulated where I am. Some are actually with me, and are simulated where you are. Some don't really exist at all, and are simulated both places.

The more of the common objects that are simulated both places, the closer you are to pure virtual reality.

We're talking, of course—that's what tea is for. The subject of robotics comes up. You ask your house for "Asimov's Robot novels" and a small bookcase appears in the air to your right. You reach out, grab, say, *I, Robot*, and hand it over. I summon my own bookcase in a similar manner and stick it in. Both bookcases disappear.

What actually happened? It was a file transfer. When I call up my bookcase next time, I'll find the book there, and I'll be able to open it and read it. It's no different, in principle, from dragging an icon across your computer screen from one folder to another.

Another form of telepresence is scale shifting. Suppose you want to build a skyscraper or nanomachine. In either case you're better off if the parts fit in your fingers' grasp easily and the whole business sits in your lap or on your workbench. With appropriately sized robotics at the other end, this is just how it would look and feel. As in the living room case, the robot could be an outsized (or nanoscale) humanoid mechanism, but in many cases it makes more sense for it to be Utility Fog, too.

Utility Fog would be nowhere near as good as special-purpose nanoengineered machines for doing any specific task. It would weigh about as much as balsa wood, be about as strong as high-density polyethylene (HDPE) plastic, and be about as hard. But it could simulate weight, strength, and hardness well enough to fool your sense of touch. It would be perfectly strong enough to simulate furniture, but it couldn't simulate a knife hard enough to actually cut something. However, it would be an extremely facile manipulator of objects, so a batch of Fog could have a stock of its own microscopic tools, like knives, to do the cutting where the simulation demanded it.

The same is true for computation, communication, and anything involving heat or chemistry (like cooking). Power and control would be provided by special-purpose units that the Fog would carry around like any other object. Individual Foglets would have simple onboard controllers and tiny fuel tanks, but there would be sand grain–sized units every few millimeters, pea-sized ones every few inches, golf ball–sized ones every few feet.

Foglets themselves would have small, roughly spherical bodies with twelve arms coming out in all directions. They could be any size above about ten microns armtip to armtip; a twelve-hundred-dots-per-inch resolution, good enough to fool your sense of touch, calls for 20 microns. If resolution to fool senses wasn't important, the robots could even be of macroscopic size. Thus we don't really need nanotechnology to start designing and experimenting with Utility Fog robots and software. But we do need nanotechnology for its ability to manufacture high-tech items in astronomical quantities. A typical room, floor to ceiling, would contain literally quadrillions of Foglets.

You can reduce this considerably, down to a few hundred trillion or so, if you're willing to give up some of the nonphysical effects such as having objects appear instantly and hover in the air. What you do is have a thick Fog carpet instead of filling the air. Then, when you want a chair, for example, the Foglets have to move physically to flow up into the shape you want. (This is called naive mode and the other is called Fog mode.) While not instant, Fog flow could easily be fast enough to form a chair in less than a second.

Fail-safe operations for the naive mode would differ from Fog mode. It would be OK to have naive-mode objects freeze into shape upon a loss of power or control signals, but in Fog mode, the air in particular would have to be able to get out of your way. If all backup power sources were exhausted, the air would retract and clump into marble or golf ball–sized spheres and leave the real air there. You'd find yourself standing in something like those play areas that are knee-deep in plastic balls that kids like so much.

Both modes have advantages. In naive mode the control is considerably simpler. In Fog mode you can simulate a barbell that would be heavier than balsa wood—the air surreptitiously pulls down on the parts you're not touching. In naive mode you don't have to generate and run a holographic simulation of the surroundings. In Fog mode you can have objects appear and disappear and levitate real objects as well as virtual ones, including yourself.

FLYING

How would you like to leap tall buildings in a single bound? Run faster than a locomotive—or pick one up? I'm sorry, you wouldn't be able to crush coal to diamonds in your hands, but hidden synthesizers could make plenty of diamonds for you.

Flying would be particularly fun. We can distinguish between virtual reality flying, where you feel like you're flying but are really sort of swimming in your cubicle, and really flying out in the open over real territory. The real form, of course, requires a lot of Utility Fog to cover the whole area. No magic—the reason you stay up in the air is that a stack of robots, albeit very tiny robots, holds you up.

Covering a vast area with a mile-thick layer of microscopic robots may seem extravagant, but I'm sure that covering Manhattan with a layer of buildings would have seemed extravagant to any of its inhabitants in 1492. It seems likely that a layer of Utility Fog would be one of the most efficient forms for the city of the future. A ten-foot cube of Fog provides enough room for virtual reality of anything for one person, and the flying ability solves the transportation problem for physical meetings within the city. The Fog supplies all the infrastructure that a current-day city has and much more.

There will likely be areas—large domes, some cities, property around houses—that are Fog-filled, but not everywhere. The advantages of filling your house with Fog we've already seen; objects can appear and disappear on command, you can accomplish most physical tasks simply by asking for them. Globs of Fog outside allow you the same luxuries in your yard, while preventing unwanted intruders—people, dogs, mosquitoes, or bacteria.

Flying could also be useful in space. Aren't you already weightless in space? Yes, but you don't have control: you can't stop and start when you want. With Fog you have control over the complete environment and your own motion. If your ship or space habitat were surrounded by Fog on the outside, it could perform tasks like maintenance, cargo stowing, and catching and docking rendezvousing craft or people.

In space, matter is scarce, recycling is paramount. Utility Fog will be one of the most useful forms of matter, since it can do so many things. What's more, being able to change from desk to bed to instrument panel to robot to bulkhead in an instant has got to count as some kind of ultimate in recycling.

REAL VIRTUALITY

One thing in a Fog world that would be more difficult than ours would be telling what was real and what wasn't. The only technology we have that's commonly used today where this is an issue is sound reproduction—good quality recordings and speakers can be difficult to distinguish from the real source of a sound.

Utility Fog will extend this to sight and touch. So will other forms of virtual reality; but Utility Fog mixes virtual and real in a way that most other forms don't.[5] What's more, with Utility Fog there's a philosophical conundrum to worry about as well. If there's a Utility Fog simulacrum of you in my living room, it clearly isn't a real person. However, I would claim that my Utility Fog couch is a real couch—nothing in the definition of a couch says it can't be made of tiny robots.

Today, we don't have any problem saying "I heard so-and-so," after speaking with him on the telephone, but we're much less likely to claim to have seen him after a video teleconference. This is not a difference between audio and video but a subtle difference in the meanings of *hear* and *see*—they aren't strictly parallel. Seeing has a stronger implication of being present. We don't notice these subtleties in normal conversation but we use them nevertheless.

The batch of Fog at the end of a telepresence link looks and acts like a person, but you know there's a real person somewhere whose actions it's just copying. What if, however, the Fog is a robot, running control programs right there on the processors of the Foglets? Surely the batch of Fog is now a "real" robot. Go one step further and suppose someone uploads his brain (see chapter 18), which is then downloaded into the Fog robot. Nothing else exists to be the "real" person now—it's the batch of Fog or nothing. It gets even worse if the download is in Fog mode, not associated with any specific group of Foglets but a moving pattern of computation.

Our language is going to have to grow to keep up with reality. It always has. We already needed this to some extent with television and computers. A picture of a bear on television is not a real bear, but *Hamlet* read on the computer screen is really *Hamlet*. The difference doesn't seem to bother anybody too much, and I'm sure that as we gain experience with the partly virtual world that is coming, we'll invent whatever new words and concepts are necessary to deal with it.

CHAPTER 14

ARTIFICIAL INTELLIGENCE
Closer Than You May Think

Ironically, in September 1984, just as the [AI] mania was at its peak, I wrote an article . . . in which I dared to be fairly pessimistic about the coming decade. And now that world has all but given up on the AI dream, I believe that artificial intelligence stands on the brink of success.

—Douglas Lenat,
Scientific American: Key Technologies for the 21st Century

Nonmobile computers are already more plentiful than robots and will always be cheaper for the same processing power, so stationary computers as smart as humans will probably arrive a bit sooner than human-level robots. What's more, stationary computers can do desk work and deal in the business or technical world without needing all the sensory and motor mechanisms that humans have for physical activities. So perhaps they will be a little easier to build, and get here just that little bit earlier.

AI techniques already pervade the business world (although often they are called other things). Every time you make a long-distance phone call or use a credit card, some computer program is analyzing the transaction and deciding whether your card is being used fraudulently. Some of these are simple, dumb, rule-of-thumb programs—I've had a card inactivated while on vacation in Australia, simply because I used it in Australia—but some are quite subtle, and better than humans at picking fraudulent from legitimate usage based on the customer's usage patterns and typical fraud usage patterns. And what's more, the systems do this for all the literally hundreds of millions of calls and card transactions that happen every day, in real time.

Think of it: some of the largest companies in the world base a substantial bulk of their sell/don't sell decisions on the word of a machine, and one following not simple rules (in the cases that work well), but sophisticated statistics-based inference procedures whose math the businessmen don't begin to understand.

When machines can do that well at other kinds of decisions, they will be put to use. Suppose similar techniques worked for hire/don't hire decisions. Or fire/don't fire decisions.

CYBERNETICS

In the 1700s, at the dawn of the industrial revolution, before steam power became the prime mover of machinery, the major sources of power for stationary mills were wind and water. Water power was straightforward to harness and generally even and reliable; but many places don't have an appropriate stream. Wind power is available many places water isn't, but there are two problems: the wind doesn't always blow at the same speed, and it doesn't always blow in the same direction.

So windmills were often built with some ingenious additions to the main spindle and sails. In 1745 Edmund Lee patented the automatic fantail,[1] a small secondary fan, set at right angles to the big one. The small fan is connected to gears that turn the whole windmill on its base. The trick is that the small fan is edge-on to the wind, and thus doesn't do any turning, except when the big fan is facing squarely into the wind. The small fan thus supplied not only the power, but the smarts, to keep the windmill properly oriented. Windmills also had automatic reefing arrangements, in which a system of weights holds the sails to the wind, but if the mill spins too fast, the weights are pulled up by centrifugal force and the sails are partially furled. Sources disagree on the origins of these, but they were in widespread use in the eighteenth century.

These devices appear to have been the origin, or at least the gateway into modern Western technology, of mechanical regulation by feedback. Feedback has become something of a buzzword in organization and communication, but before that, it was a buzzword in engineering. In the engineering sense, it simply means a process in which the output of a machine is measured, and the measurement used to control the machine.

One of the simplest and most common examples of feedback in control is the thermostat. A typical home heating thermostat is a combination of a thermometer and a switch that turns the furnace on or off. As the temperature in the house rises above the desired temperature, called the setpoint, the thermostat turns the furnace off; as it falls below the setpoint, it turns the furnace on.

There is some lag in this system, producing what's called hysteresis. First, the thermostat itself doesn't actually turn the furnace on and off at the same temperature. For example, it may turn it off as the temperature rises above 72, and only turn it on again as it falls below 70. In addition, it may take some time for the heat to seep out from the radiators across the room to the thermostat. As a result of both of these, the furnace doesn't just flicker on and off but stays on or off for longer periods of time, which is good for the furnace but causes the temperature to swing up and down.

Hold an arm out at full length, and stick a finger out sideways so you can see it. Now watch the finger with reference to some fixed point in the background behind it, and you'll find that it isn't actually rock steady. It wanders around the point you're trying to hold it, in a way very similar to, and for the same reason as, the temperature fluctuates around the setpoint of the thermostat. In other words, your own control of your muscles uses feedback.

Let's assume for the sake of argument that you would rather have the temperature (or the arm) be steady. One thing you can do is get a furnace (or circulator pump or fan) that isn't simply on or off, but which can be set to run faster or slower. The problem now is that the thermostat has a harder job to do. It isn't enough to turn on or off—it has to tell the furnace how hard to run. (Your nervous system already does this, of course—you don't hold an arm in position by alternately jerking your muscles absolutely tight and then going completely limp.)

It would help if we knew the speed to run the furnace that kept the house at an even temperature. Then the thermostat could have the furnace run at that speed at the setpoint, and run harder if it was colder, to warm up to the setpoint, and run slower if it were warmer, so as to cool down. Such a scheme is called proportional control. Note both of the windmill control schemes described above were examples of proportional control—more sophisticated than many modern-day thermostats!

The problem with proportional control is not only that you don't know in advance what that magic even-temperature speed is, but that it changes. Changing temperature outside, opening of doors or windows, and/or other heat sources inside can all throw off even a correct initial estimate. For example, if someone's left a window open, you need to run harder to maintain the same temperature. And indeed, a simple proportional control will allow the temperature to vary somewhat with the conditions.

A smarter system might have a thermometer outside as well as inside, to let the thermostat guess how hard it needs to run the furnace. We can imagine sensors on the windows and doors, and sensors to tell how brightly the sun is shining and whether the roof is snow covered. In reality, of course, houses don't need nearly as sophisticated a control, but some things do. Animals—and humans—are a case in point.

Typical home heating systems aren't set up so that the room thermostat controls the furnace; it controls a circulating pump, and there's a thermostat on the boiler that actually controls the furnace. It would be possible, although it's not common, to have a proportional control of one level of such a duplex system control the setpoint of the other. It's not common in house heating, that is; in more complex systems, it's much more frequent. The nervous system of an animal, or even an insect, is a complex hierarchy of feedback loops controlling feedback loops.

It's clear that the character of nervous systems as feedback control systems was one of the prime motivators for cybernetics. Cybernetics was the original mathematical treatment of feedback for control in higher math, done by Norbert Wiener in the World War II timeframe.[2] It was sucessfully used in the design of automatic radar-controlled antiaircraft guns, among other things. There were high hopes that the theory could be extended to form a basis for understanding everything from animals to economies. And in fact, dynamic feedback loops do pervade a wide range of phenomena. But the mathematics did not, in the end, suffice to put a broad, common base under all of them, and the term *cybernetics* and many of its grander dreams faded away.

Meanwhile, control theory, as a more restricted branch of engineering, flourished and is now considered one of the fundamentals of the field.[3] Feedback is extensively used in mechanical, hydraulic, and electronic systems, and its analysis is not considered anything other than part of the engineer's standard toolkit. It can be used to design active creatures of considerable complexity. It is clearly present all throughout the structure of actual nervous systems. The broad range of mental phenomena such as hypnosis, trances, dreaming, and the variations of consciousness ranging from intense concentration to woolgathering speaks tellingly of an organization much more like a network of feedback loops than a straightforward algorithm.

GOOD OLD-FASHIONED AI

The history of science is inseparable from the notion of reductionism. Take a complex natural phenomenon—the classic one is the courses of the stars and planets in the heavens. Reduce it to a simple model—in the case of astronomy, the model was ultimately a very few numbers for each planet, denoting its mass and position—and simple rules (often in the form of differential equations) for manipulating the models.

As artificial intelligence (AI) got its start, naturally an attempt was made to do the same thing that had worked so well for the physical sciences. A classic example was language translation. A very simplistic model of translating from one human language to another is to find the word in the target language that corresponds to each word of the source-language text and string them together to form the output text.

As you might expect, such an approach doesn't work very well. The results are perhaps best characterized by the jokes that come down to us from that period: "Some researchers decided to test their translating computer. They gave it the sentence 'The spirit is willing, but the flesh is weak' to translate into Russian, and they they had it translate the result back into English. What the computer told them at the end was, 'The vodka is strong, but the meat is rotten.'"[4] (Author's aside: The free translators one finds on the Web today aren't much better; I was able to get one to produce essentially the same result, going through Portuguese instead of Russian!)

Oh well—back to the drawing board. Some phenomena are

harder to model than others, after all. Physics took a bit longer to handle electromagnetics, or fluid flow, than gravitational dynamics at the planetary scale.

So AI researchers went on to other things; and some really impressive results began to appear. AI programs at the complexity of one PhD thesis began to be able to do things like solve freshman calculus problems, play chess, or do analogy problems like those found in IQ tests.

Perhaps the crowning achievement of this period was Terry Winograd's SHRDLU program,[5] which was capable of playing with (simulated) blocks, holding a conversation about what it was doing, and to all appearances really understanding what its correspondent was saying in simple conversational English.

SHRDLU was a conceptual breakthrough because it contained a semantic model—an inner program that could be used as a simulator in a way that gave meaning, if only the simplistic glimmerings of it, to the symbols the program was manipulating. In other words, SHRDLU could seriously be said to understand the things it talked about, if only in a very limited way.

What it couldn't do is learn (except by being given an explicit definition for a new word). This takes nothing away from SHRDLU—it was every bit the milestone I described, and it was written in a day when all the computers at MIT together couldn't match one of today's PCs for raw bit-flipping power and storage capacity.

The problem is that in the thirty years since SHRDLU, no one has come along and finished the job. The things SHRDLU demonstrated—parsing, disambiguation based on a prefabricated semantic world model, manipulation of such a model, and so forth—have been improved almost beyond recognition. We have chess-playing machines that can best any human, but life isn't chess. We have programs that do amazing searches in amounts of data much bigger than any human could handle. We have systems that are indispensable for the design of ever-more-complicated machines, and that facilitate the design of ever-more-complex software. At the experimental level, computers can converse in spoken English, recognize faces seen with video cameras, and control robot bodies that walk on two feet.

The AI community has taken a somewhat justified stance in response to its critics, that whatever it was that computers couldn't do was considered AI, while whatever they succeeded in reducing to practice was promptly moved to the field of "known software technique" and not considered AI anymore. So AI was never given credit for what it accomplished, but was always measured on what it hadn't done.

After all, in the 1960s, playing chess and doing symbolic algebra was considered AI. In the 1970s, image and natural language processing were considered AI; in the 1980s, inferential databases and rule-based diagnostic and configuration systems were considered AI; in the 1990s, robots that drove cars or walked were considered AI.

On the other hand, in the 1950s, translating languages was considered AI, and it still is. And to be fair in the other direction, AI has generally defined itself in such a way that it would naturally be measured by the difference between what computers can do and what humans can. So how far has it gotten? The things that have been achieved form an impressive list; but they are a small fraction of the things that humans can do. Let's take a look at the things that computers still can't do.

In the field of vision, for example, it's now fairly well understood how to duplicate the preprocessing that the few layers of nerve cells in the retina do to the visual image before sending it along the optic nerve to the brain. This may sound trivial, but it requires one thousand times the processing power available to the early AI researchers!

To continue, the reverse engineering of the visual system has proceeded to the point of being able to separate and distinguish objects in the visual field, and place them in a three-dimensional world to some extent; but the current state of the art doesn't seem capable of general identification, that is, telling you what each object is.

The current-day descendants of SHRDLU can converse in colloquial English about specific, well-defined topics, and current-day expert systems do professional-quality jobs on some impressively complex and subtle tasks.[6] But if you get these systems out of their field of expertise, they don't have the ability to learn the new stuff or even react with common sense and general intelligence to what they experience.

In the field of robotics, the state of the art ranges from assembly manipulators in factories, which do simple manipulations at blinding speeds, to Asimo, the famous Honda robot in the shape of a human, which walks, albeit slowly. In between are the autonomous soccer teams (of small, wheeled robots) that play respectable games of table-top soccer (where speed and strategy are of equal importance).

And yet, present-day robots are still very, very primitive by the standards of animals, or even insects. Rodney Brooks, head of MIT's AI Lab and famous for creating several of the revolutionary robots of the 1980s and 1990s, goes so far as to say that we're still missing some basic key concept, in the way we were missing calculus while attempting to understand the motions of the planets, before Newton and Leibniz filled us in on it.[7]

(It's just a little bit ironic that Brooks made his name, and to some extent revolutionized robotics, by tossing out the mechanisms that AI had always assumed needed to be there, and building robots that ran on reflexes rather than logic.)

However, it doesn't seem to be the higher reflective functions that Brooks is worried about; in fact, when he characterizes his notion of what's lacking, he holds open the possibility that we may just need more parameter twiddling, or more computer power, or any other normal, expected progress in the field. Still, he seems to think that what's needed is that a lightbulb should go off in some researcher's head and a "Eureka!" moment would then occur.

When you watch the actual robots or interact with the programs, it's not easy to decide which side of this question to agree with. On the one hand, they are enormously, almost incredibly, better than they were a quarter century ago. And yet, they are still so awkward, "lumpen," or clueless that there seems a difference in kind, not degree, between them and the natural systems we're trying to emulate.

Perhaps one way to characterize what the difference feels like is that it's like comparing a pencil line drawing to a color photograph. The drawing can be enormously suggestive of the picture, clearly revealing a depth of perception and skill on the part of the artist, as the AI programs do about their creators. Yet there are a thousand subleties the drawing doesn't capture that are evident in the picture.

Part of this is clearly due to the fact that AI systems don't have the computational power to throw around that living ones do; brains still outperform silicon, ounce for ounce and watt for watt, by a few orders of magnitude. But this is, and has been, changing, and it's very likely that that particular constraint will be gone inside two decades.

Another reason for the parsimony of artificial systems is more endemic to their nature; it's the same reason the pencil drawing is more parsimonious than the photo. The drawing was done one mark at a time by a person; each detail, and its relation to each other detail, occupied his consciousness at some point. The more detailed the drawing, the more effort was required in its construction. The same is true of AI programs. (Yes, a multiperson development team, as is used for any significant commercial software, does help. This is still not hugely prevalent in AI so far, but as AI moves out of the academic into the commercial world it will increase.)

WHY AI IS HARD

As 1999 turned to 2000, people across the world gathered to celebrate the new year, the new century, the new millennium. Once the parties were over, some wags, including your humble narrator, pointed out that there was still a year to go and that we'd have to do it all over again. Centuries begin with the "01" years because there wasn't a year 0, and so the first century began with year 1.

Numbers can be used in two subtly different ways. First there is counting, to tell how many of something you have. But there is also naming. If, for example, I live at 6742 Waycaster Street, it doesn't mean that there are 6742 houses from one end of the street to mine. The number 6742 acts like a name. And it would still act like a name if there were no gaps; the first house would be named 1, the second 2, and so forth.

Look at a ruler. You know there are twelve inches on it. Yet typically, there are either eleven marks dividing the inches, or thirteen counting the ends; not exactly twelve. The numbers are measurements denoting the points between the inches, not names of the inches (in which case, of course, there would be exactly twelve). Yet the year numbers are used as names of the years: hence the confusion.

When you're programming a computer, you have to understand exactly how you're using your numbers or you'll introduce bugs into your program. (Bugs arising from this particular confusion are sometimes called fencepost errors.) And yet most people manage to live their lives happily without ever being aware of the distinction, much less applying it correctly in each situation. And what's more, we usually get it right anyway—we go to the right house, using the number as a name, and measure the room, using numbers as measures, and don't even notice the difference. We have a concept of counting that includes both, and somehow it seems to work.

This couldn't possibly work if you were built the way current software is. It's built from the bottom up, and all those details have to have been noticed, and gotten right, or the program doesn't work at all. We humans seem to be able to come at things from the top down, making finer distinctions if necessary but getting by with no trouble on general notions in most cases. Conventional software, such as the applications on your PC, just runs in the opposite direction from the start. Literally millions of tiny details, each critical to get exactly right, combine to give the overall effect that you see.

Besides coming at the problem from the wrong direction, there is the business of the amount of detail described above. Because software is built from the bottom up, building an AI program is much more like building the brain a neuron at a time than like starting from some condensed, high-level description such as the genome. Viewed from this perspective, a mature human brain is much, much more complex than the high-level description—perhaps a million times more complex. Where did that complexity come from?

The best way to describe it is that the brain is autogenous: it grows in patterns that depend on both the original instructions and on the inputs from the senses. It grows, it learns, it builds itself.

In other words, the nature of the code in AI programs is different in a fundamental way from the nature of the code in the genome that builds the brain (even in the highest, most abstract, analogy): the AI is a program for doing cognitive tasks, but the genome is a program for building a machine that does cognitive tasks. Or perhaps for building a machine that builds machines to do cognitive tasks.

NONTRADITIONAL AI

While symbolic AI was taking off in the 1960s, a number of biologically inspired alternatives got left by the wayside. Besides cybernetics, they included neural networks and genetic algorithms.[8] When purely symbolic systems began bumping into ceilings in the 1980s, the other approaches enjoyed a new wave of popularity and investigation. Given considerable new scope by the processing power that began to become available by the 1990s, these methods made great strides.[9]

The specifics of these methods are not important to this discussion; there are many of them, each with lots of variations. What they have in common is that they lack the brittleness of the rule-based approaches, but they also lack the ability to handle large complex problems. One of the features of the nontraditional approaches is often that they learn the function they are supposed to perform, an advantage, but by the same token they require masses of training data, which can be a drawback.

It often turns out, by the way, that the methods derived from biological (and physical, and other) analogies, which work, can be analyzed so as to reduce the model to something more straightforwardly statistical. For example, neural networks can be trained much more efficiently with an ODE (ordinary differential equation) solver than the biologically inspired "back-propagation" algorithm.[10] Methods such as Bayesian occupancy grids for robot navigation simply adopt the statistical basis to begin with (though perhaps inspired by neural maps), but have much the same feel, with many of the same strengths and weaknesses.[11]

What is needed, of course (and this has been obvious to AI practitioners for decades), is to combine the techniques in such a way as to add the flexibility and learning ability of the biologically inspired approaches to the structured complexity of the symbolic ones. Many attempts have been made to do this in various ways.

One of the more interesting ones was the series of programs produced by Douglas Hofstadter's group at Indiana University. The most famous was Copycat, by Melanie Mitchell.[12] Copycat was a program that answered letter-string analogy problems. (For example, if ABC becomes ABD, what does PQR become?) The goal of Copycat was to explore several techniques that differed from straightforward algorithmic, symbolic ones in favor of flexibility ("fluidity," in the group's phrasing).

The essence of what Copycat did was to build interpretive structure, like a classical symbolic AI program, but using a probabilistic method based on a biological model: the inner workings of a cell. In other words, Copycat worked internally like a software version of diffusive transport and self-assembly. Thus Copycat did have, within its microdomain, the combination of flexibility with symbolic complexity that AI needs.

Copycat, like SHRDLU, illuminates a significant building block of cognition. In SHRDLU's case it was understanding as a function of a model. In Copycat's case it is a general operation we can call analogical quadrature. Copycat's problems are like the analogy problems you often see on IQ tests. The point is to make an analogy between two things and carry the analogy through a transformation of one to transform the other appropriately. We can imagine, though no one has written so far, a scaled-up industrial-strength version of Copycat that could do analogical quadrature with the same insight, flexibility, and occasional inventiveness that Copycat itself brought to letter sequences, but do it to data structures in a full-fledged modeling language able to describe objects, situations, actions, and events.

Another similarity Copycat has with SHRDLU, unfortunately, is that it doesn't learn its capabilities; its total stock of knowledge, carefully tuned and coded, is set up by the human author. Again, this in no way detracts from what it does do; this is just to remind you that many deep problems remain to be solved.

A SYMBOLIC SERVO

Back in the dawning days of AI, two of the founding fathers, Allen Newell and Herbert Simon, created one of the first landmark programs, called GPS (meaning General Problem Solver). GPS worked by being given a description (in a high-level, symbolic language of course) of a current state and a goal, and having a list of possible actions, it would choose the one that would most reduce the difference between the state and goal, and continue, starting with the new state. The story is told that when AI founding father John McCarthy first saw GPS, he exclaimed, "Why, it's a symbolic servo!" (To complete the AI pantheon, I'll mention that it is Marvin Minsky who tells the story.)[13] A servo is a feedback control loop as described in connection with cybernetics above. GPS works, as well as any symbolic AI system does—its descendants include Soar, the most successful general-purpose classic AI system to date. It is clear that lower levels of the nervous system act like conventional servos. It's also clear that among the other things going on in the upper reaches of cognition, something like GPS's (or Soar's) symbolic manipulation is happening. What if we could find some general mechanism that would bridge the gap? Then we would be able to think about a mental architecture that was a complex hierarchy of feedback loops, as cybernetics and neurophysiology suggest, but retain the ability to do the symbolic stuff that has been the weakness of that level of theory.

Here's a possible candidate. It's based on analogical quadrature and an "industrial-strength Copycat." It's called a sigma servo or sigma for short, which stands for SItuation, Goal, Memory, Action. Imagine it's embodied in some small chunk of the brain (which consists of many, many such chunks, complexly interconnected). It has as inputs sets of sensory signals, possibly connected directly to sensory organs, possibly highly interpreted, but in any case signals that may be taken to define a situation. It has nerve signals coming from above in the hierarchy, which can be thought of as a goal. It has outgoing signals it can generate that are thought of as its actions. And it has a memory of its past activity, which is organized as triples of situations, actions, and the new situation that followed after performing the action, which we will call the result.

The sigma now does an analogical quadrature

[situation, result in memory] is to [corresponding action in memory]

as

[current situation, goal] is to [new action]

between its current inputs and every remembered triple. It picks the one that has the best overall match (or interpolates between multiple close misses) and does the new action. (In a somewhat more developed version, there is an aversive flip side, where results are remembered as painful, and corresponding actions suppressed, as well.)

The level at which the sigma servo operates is now completely general. It has its own memory, written in its own language. If situations and actions are simple numbers, quadrature works out to be interpolation, and the servo is a proportional controller. This is typically as sophisticated as is needed for, say, controlling a single muscle. In this case, as with the higher-level case in general, the situation is augmented with a history-gathering and -remembering mechanism.

Note that at the lower levels, it can be continuous in operation. As the input signal varies over time, the output varies, too. The higher and more symbolic it gets, the more discontinuous jumps there must be, until you get to something more like a finite state machine. Note that two of today's leading roboticists, Albus and Brooks, favor cognitive architectures that are hierarchical feedback networks of finite state machines.

With highly abstract, structured, symbolic memories, the sigma servo becomes a processing engine of the AI sort. Indeed, the situation begins to look a lot like what Minsky called a frame.[14] And variations in the kinds of signals processed and remembered allow the servo to be adapted to a wide variety of tasks. More to the point, there's not just one of them, but a complex network, and the situation for some is a description of what the others are doing—and their actions consist of setting the others' goals and/or characterizing each other's results.

One last point: the triples in memory, particularly at higher levels, don't come just from one's personal experience but from watching others (quadrature translates them to apply to oneself), hearing stories (the best stories are chains of causal triples about a character with whom the listener can identify), or even running models in the imagination, "letting our theories die in our stead," in Karl Popper's marvelous phrase.

THE POWER OF THOUGHT

Wait a minute, you might say. The sigma does this complex operation, equivalent to Mitchell's entire PhD thesis program, to every single memory at once? Doesn't that let us in for a lot of useless computation? (By the way, when this use-all-the-memories technique, or something like it, is used in AI, it's called case-based reasoning.)

For some cases, yes; but for others, that's the cost of generality and flexibility. Beyond that, I should point out that the brain is not like a von Neumann computer architecture with an active processor on the one hand and a passive memory acting like a filing cabinet on the other—the memories are right there in the neurons that are doing the processing, stored in what form exactly we don't know yet, but the short answer is, "It's all processor." What's more, once we know that the memories are going to be matched in a certain way, it's not a particularly difficult feat of programming to index them so that the bulk of the work of matching is done on the ones closest to a reasonable match.

The other, more interesting part of the answer is that putting memories in isn't like stuffing paper into a filing cabinet, either. Memories fade with age, unless reinforced. Memories similar enough

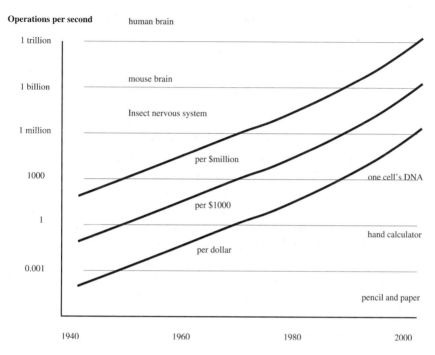

Figure 23. Moore's Law. For maybe $5 million, maybe $50 million, human brain computing equivalence is available now. The software still needs some work.

merge into generalities.[15] (Note that since there are many sigmas specialized for many purposes, memories will be merged and generalized in some but not in others. The ones in your balance control have merged all the steps you've ever taken into a nice smooth control model, but somewhere else, the steps up onto the podium to receive your diploma stand alone.)

This is, of course, where nanotechnology and Moore's Law come in. The processing power of the brain has been estimated at a million operations per second per neuron times ten billion neurons. Using fancy programming tricks we might cut that by a factor of one hundred to one thousand, but probably not much more. That leaves us needing at least a thousand, possibly ten thousand, current-day high-end PCs to run a full-fledged AI.[16] Now we have at least one idea about what all that computation might be doing.

It's worth pointing out that some scientific and commercial computer systems have this kind of processing power right now. Of course, they cost tens of millions of dollars. Moore's Law puts this on your desktop for one thousand dollars in twenty-five years.

Don't expect to wait anywhere near that long for AI to show up, though. If a top engineer or manager could be replaced by a $10 million machine that worked twenty-four hours a day and never took vacation, at today's salaries, it would be economical to do so. The hardware is here now; what we lack is the program. Considering the general rate of progress in the field, I'd be quite surprised if broadly human-equivalent performance weren't available in ten years or so. At that point, today's $10 million system will cost $175,000.

So suppose you sell your house and buy one? What could you do with it? If you have an information job, anything from architecture to managing a store, the machine could do it for you (with a few cameras and speakers in the right places). Suppose your job is writing AI software? Your AI proceeds to write a better version of itself, with no further help from you. Better can mean two things, and usually means both: it can do more or be faster with the same hardware or it can take advantage of more hardware to do even more than that. In either case it's now smarter than you are.

SINGULARITY

The reciprocal of one-tenth is ten. The reciprocal of one-hundredth is a hundred, and the reciprocal of one-thousandth is a thousand. As the number gets closer to zero, the reciprocal rises precipitously. Zero has no reciprocal at all; zero is called a singularity point for reciprocals.

Vernor Vinge, computer scientist and a grandmaster of science fiction, coined the term *singularity* to describe a point in the future where the intelligence of machines increases precipitously.[17] It's not likely that it will be a true mathematical singularity, which would imply that progress would increase without bound before a specific date. However, as a general term for a time in the future where machine intelligence increases to the point that we humans are unable to keep up, it has caught on as a buzzword among the technophilic avant-garde.

First of all, can this really happen? There are alternate theories:

suppose that human intelligence is an optimum point and that adding more processing power hits a law of rapidly diminishing returns. Committees certainly give credence to this idea. It's true that we don't know how to organize a superhuman intelligence, or even a human one; but rapid strides are being made in the area, particularly in understanding how human and animal minds operate. It seems likely that we can do at least a little better once we have a full understanding.

Another indication that AIs could exceed humans is that AIs could have low-level, instinctive access to the kinds of abilities that computers have now, but that humans do poorly, such as solving partial differential equations, maintaining billion-fact databases, and counting the words in this book in approximately a millisecond.

So are we to be outclassed, left in the dust, adopted as pets if we're lucky, and in general suffer a crushing blow to the human spirit because we're no longer the smartest kids on the block?

Of course not. For one thing, we built these machines and we get proud-grandparent credit for anything they can do. (Let's make sure it's credit and not blame we're due!) And then, we always have the option of improving ourselves as well as the machines, something we'd want to do anyway.

But the most basic reason is that most of us aren't the smartest kid on the block to begin with. There's only one per block. Most of us aren't haunted by nightmares because we can't beat Deep Blue at chess; we can't beat Garry Kasparov, either. And none of us can beat City Hall.

Synthetic organizations—clubs, societies, guilds, and institutions—as distinct from autocracies and natural, fluid groups of people, have been around for thousands of years, and nothing is a clearer trend in history than the rapid rise in their size and complexity. It seems a good bet that the first general AIs will show up in the decision-making apparatus of corporations, since that's where the need is great and the resources are available. Working for, or trading with, such a machine won't be all that different from dealing with the current paper machine. When it is different, it will in general be better. Corporations want their customers to be satisfied, and they want their employees to be happy—or they'll lose unhappy talent to ones that do.

Another crucial implication of AI is that the design of ever-more-complex machines and systems will continue to be possible. We should expect that we'll be able to have systems more complex than we do now, which work better and are more reliable, and will still be able to afford them.

Suppose that in 2010 a certain system costs $10 million to design. Assume that's when the human and AI designers are equivalent in cost, and the AI designer gets cheaper at Moore's Law rates. Then in 2020 the same system would cost $175,000 to design; by 2030, $3,000, and by 2050, less than $1. Somewhere in between it gets economical to design each individual item, not just for the individual person using it but for the particular instance of use. Not just customized—a whole new kind of machine invented as if a top-notch engineering team spent years working on it, just for your task of the moment.

This would be impossible to cope with if you had to learn a whole new set of techniques, commands, and capabilities each time. But the new device will contain an AI as well as being built by one. It will be an expert at the kind of job it is supposed to do, but not only that; it will know you, understand your style of operation and interaction, and be familiar with the interfaces you are used to.

So when you go to the garage to go to Phoenix alone, you find a completely different kind of car waiting than if the whole family were going to Nova Scotia. It might seem overkill to reinvent a whole new kind of vehicle for each trip, but you can see how something like this will come in handy for controlling Utility Fog.

CHAPTER 15

RUNAWAY REPLICATORS
Playing with Fire

Arthur remained very worried.
"But can we trust him?" he said.
"Myself I'd trust him to the end of the Earth," said Ford.
"Oh yes," said Arthur, "and how far's that?"
"About twelve minutes away," said Ford, "come on, I need a drink."

—Douglas Adams,
The Hitchhiker's Guide to the Galaxy

Sometime in the twenty-first century, someone will release an aerovore into the Earth's atmosphere. Note I don't say "may" release—I feel reasonably certain it will happen, no matter what precautions we take against it.

Most people's reaction to a statement of this sort is, "What's an aerovore?"

An aerovore is a form of gray goo, a self-replicating nanomachine that feeds on the natural environment and exists only to convert as much of the universe as possible into copies of itself.[1] The term *aerovore* means a creature that eats air.

Living off thin air is hard work. Life flourishes almost everywhere on Earth, living on rocks and deep sea vents and so forth. No natural life-form (to the author's knowledge) lives on air alone, although the air is full of spores and microorganisms going from one place to another. Nanotechnology makes it possible, however, to live on just air and sunlight.

An aerovore would work something like this: It floats in the air like a particle of smoke. Its structure is mostly carbon, which it separates out of the carbon dioxide (CO_2) in the air. It needs energy to do the separation, so it has solar collectors. The solar collectors are like antennas that stick out the sides, making it look something like a miniature black snowflake. (It would be too small to see with the naked eye.) It would be black to catch the sunlight; some would be converted into power, the rest would warm up the air around it, providing lift. (The aerovore might also use, or consist of, a hydrogen-filled balloon.)

Such a device might be created as a doomsday weapon, for example. The intent would be to have aerovores multiply throughout the atmosphere, causing a nuclear winter effect. Alternatively, an environmentalist group might try to use them to undo the greenhouse

Figure 24. An aerovore. Feeding simply on air and sunlight, an aerovore begins to make a copy of itself.

effect and stop global warming. We have plenty of evidence that people with either of these motivations can command significant resources and have no compunctions against acting unilaterally.

What will happen when someone releases one? Probably nothing much. An aerovore must be a full-fledged Stage V nanomechanism, highly optimized. It's trying to live in a niche that natural life can't manage to exploit, after all. It will be able to carry little radiation protection, so it will need to be small and fast to reproduce before it gets zapped by cosmic rays. If it lives very high in the atmosphere, it gets exposed to more radiation. If it lives low, it gets rained out. The aerovore's design is a dense web of engineering trade-offs. It needs lots of power to separate carbon and hydrogen, so it must have plenty of solar collectors. It needs to sort through about three thousand oxygen, nitrogen, and argon molecules for every usable one of CO_2, so it must have many molecule sorters. It needs to build offspring as fast as possible, so it must be packed with construction machinery. All these needs conflict, and the chances are aerovores won't multiply very well if they aren't balanced just so.

What's more, building nanomachines purely out of just carbon, hydrogen, oxygen, and nitrogen (CHON for short) is a tough constraint. All of the working parts seen in simulation so far use other elements, ranging from sulfur to flourine, to have a better-stocked toolkit of atomic sizes to get the shapes you need. There's a good chance we're going to need metals for various mechanisms. The molecular machines in your cells need a wide variety of metals in trace amounts. An example is the iron in hemoglobin, which gives a favorable electron distribution to an oxygen-catching pocket in the molecule. The aerovore needs oxygen-catching pockets in its molecular sorting rotors.

Let's assume, however, that the engineering has been gotten right and someone builds an aerovore that works. Then what? Do they block out the sun?

The first thing to note is that there isn't all that much CO_2 in the air to begin with. Let's assume that the aerovore is a cubic micron and contains ten billion carbon atoms. Building the aerovore compacts the carbon from its original form, where it occupies one part in three thousand of the air, to one part in a million. (These are all parts by volume—parts by weight remain unchanged.) It works out to about a thousand aerovores per cubic millimeter of air.

Unfortunately, the collectors increase the effective diameter of the aerovore to, say, 5 microns. If they convert all the CO_2, the aerovores in one cubic meter of air would have an effective surface area of four hundred square meters. Not much sunlight would get through. The Earth, seen from space, would soon look dark, instead of blue-white. Aerovores killed by radiation would fall to the surface as a fine black snow. Vegetation would begin to die.

Of course, aerovores are also a form of vegetation, artificial of course, but dependent on sunlight no less than trees or grass. What's more, they would rapidly eat out their foodstock, the CO_2 in the air, at the highest level they could operate optimally (lower levels would be shaded by the higher levels). With low CO_2 levels, aerovores would not be able to replicate fast enough to beat the effects of radiation and would begin to die off. In the long run, they might simply starve themselves out of the air. No one has done a serious climatological simulation yet, since the parameters of aerovores are not really foreseeable at this point. But it seems possible that they could deplete the CO_2 and largely die off, plunging the Earth into an ice age. Or an ice age might be avoided, since black aerovore snow might prevent glaciers from reflecting so much sunlight back into space.

What's a planet to do? Nuclear winters, uninvited ice ages, smudgey skies—all are best avoided if possible. One obvious thing is that aerovores, like any form of vegetation, represent a resource. They, like green plants, have already done the hard work, and we need only harvest the fuel their carbon represents in concentrated form. Rather than digging coal mines in the ground, we need only fly through the air with big filters and catch the stuff. A nanoengineered reaper could look like a gauze veil the size of a football field, and float through the air sweeping the sky. (It would be essentially invisible from the ground.) Such a skysweeper would burn part of its catch for its own energy and send the rest down to fuel collection depots on the ground. Unlike aerovores, skysweepers could operate at night.

As an aside, aerovore smoke at its densest could fuel an airplane flying through it, but we wouldn't want to let it get that dense—all the bad effects mentioned above would occur.

Skysweepers are a good idea whether aerovores are around or not. The sky is full of all sorts of gunk. You may be suprised to learn that about 10% less sunlight reaches the ground today than in 1950.[2] Some of the haze causing this is from industrial effluents, but more is from biomass burning, for example, wood and dung, that forms a major energy source in the (densely populated) unindustrialized countries. Wherever the gunk comes from, the air could stand to be cleaned up. If we did clean up the atmosphere, we could tolerate a 10% blockage owing to aerovores and still get the same amount of sunlight we do today. Better yet would be to keep it completely clean!

BIOVOROUS NANOREPLICATORS

The gray goo scenario is more often imagined as a form of mechanical bacteria that live on the biosphere. It's commonly imagined by thriller novelists that such nanoreplicators could be much more voracious than natural bacteria, reducing the biosphere to gray goo (hence the name) in the timeframe of a major motion picture.

On the other hand, other commentators pooh-pooh the idea that such replicators are even possible.[3] Apparently they haven't heard of bacteria. But of course the existence of bacteria doesn't prove the possibility of something significantly more efficient than bacteria to pose a threat.

Birds fly. We can build flying machines that, operating on basically the same principles, are faster. Yet our airplanes are distinctly inferior to birds in some very important respects. They need a substantial support infrastructure; birds can operate anywhere. Airplanes need special highly refined fuels and are built of special, highly processed materials; birds get both from the natural environment by their own efforts. Birds repair themselves and reproduce themselves. If you tried to build an airplane that drilled and refined its own fuel, mined, smelted, and shaped its own aluminum, contained its own repair shop, and manufactured new airplanes as well, it's not so clear that it would fly faster than a bird. In fact, it probably couldn't get off the ground at all.

But bacteria also build themselves out of found materials and use them for fuel. They grow, repair themselves, and replicate. And they do it with remarkably high efficiency.

Nanomachines, on the other hand, will be more like the airplane. They'll be made of materials requiring a lot of processing and use specially prepared fuels and feedstocks. They'll use higher-energy reactions and operate at higher speeds than life's machanisms. They'll be faster than bacteria, just as airplanes are faster than birds. But they'll have the same drawbacks as well, and for the same reasons.

A novelist can put birds and airplanes together and imagine a creature that lives on worms and flies at the speed of sound. Novelists can also put bacteria and nanomachines together and imagine super-fast nanoreplicators that live on anything and eat the environment out from under us. But in both cases, there just ain't no such animal.

THE OSTRICH STRATEGY

It is certain that some people reading this prediction will have the immediate reaction that it would be better if nanotechnology were never developed at all. Some people are making public cases for this idea already. They would like a law to be passed that would prevent nanotechnology from being possible.

Their like has been seen before. In 1897, House Bill #246 in the Indiana House of Representatives proclaimed that the value of pi would thenceforth be not its actual irrational value, but one of three different rational values depending on whether you were trying to find the circumference of a circle, its area, or do trigonometry. Luckily for Hoosiers, the bill died in the state senate.[4]

It seems likely that similar things will be attempted for nanotechnology. Current-day legislators are no more scientifically astute than their nineteenth-century counterparts. For example, Caroline Lucas, the Green Party Member of the European Parliament for southeast England, proclaimed, "The commercial value of nanotech stems from the simple fact that the laws of physics don't apply at the molecular level."[5]

Activists and politicians somewhat more in contact with reality will most likely opt for laws that, while implicitly admitting that nanotechnology is possible, will attempt to prevent people from developing or deploying it. That is, they will first stop people who are currently claiming to work on nanotechnology. This will bring a great buzzword shift, as all the people who had been working on chemistry, or surface physics, or coatings, or material science, and began calling it nanotechnology in the past decade will go back to calling it chemistry, or surface physics, or coatings, or materials science. Next, the laws will have to clamp down on all those people and somehow erase all the knowledge they've generated. Similarly, biotechnology, one of the most direct pathways to nanotechnology, will have to be stopped: not just prevented from selling products, but the knowledge curtailed. Finally, you'd have to outlaw high-precision machining and watch-making, since the original pathway proposed by Feynman lay in that direction. Indeed, you'd have to stop most technological progress in its tracks, since any improvement in precision, control, or chemical synthesis is a step toward nanotechnology.

Well look, they will say. We did it with nuclear technology. Since World War II, nuclear technology has been "born secret" in this country. Under the strictest regulation possible—there is a death penalty, in some cases, for sharing scientific knowledge, which has been enforced—the progress of nuclear technology has been effectively quashed. Compared with other technologies in the latter twentieth century, nuclear has been at a standstill. Worldwide treaties prevent proliferation, and any development of nuclear technology in heretofore nonnuclear nations is viewed with concern.

And yet proliferation has occurred anyway. It is important to point out that nuclear technology is quite difficult. Whatever path you take has to go through uranium. Uranium is scarce, difficult to work with, easy to detect—and you need tons of it. Even with no restrictions, nuclear technology would be the province of governments and large corporations.

Nanotechnology, on the other hand, can be done, if you follow the biotech path, with bacteria that are literally sitting on the end of your nose. You can choose any of hundreds of pathways. The size of an effort to develop it, given current knowledge, is probably about that of the science-oriented departments of a typical state university, and getting smaller every day.

What's more, nanotechnology has a lot bigger market than nuclear weapons. Again, only governments can afford to own nukes, but almost everybody has a use for nanotech products. Cancer cure, anyone? Life extension? Intelligence enhancers, fat reducers, AIDS prevention?

Another major difference is that nuclear technology was throttled without strangling a lot of existing, economically important, fields. Things nuclear are a world apart from the operations of ordinary matter. But life itself—medicine, understanding ourselves and our environment—is one of the major pathways to nanotechnology.

Mechanical engineering is another. Catalytic chemistry, materials science, microchip fabrication: virtually all the technical background of standard products today forms a thicket through which a host of pathways could be picked.

If general laws are passed to impede nanotech development, the military in the United States and almost every other country will be exempt and develop it anyway. So will organized crime, various international drug cartels, and probably some environmental groups.[6] Luddite movements will attempt to demonize nanotechnology as well as other technologies, but people will start smelling a rat when senators continue to hold office past age 120.

Short of plunging the whole world into a Dark Ages, you simply can't hide the laws of physics. The basic physics necessary for nanotechnology has been in the textbooks for decades. The ideas are out there, the pathways are many. The more that legitimate, open, advanced technology is restricted, the more people will look for other options.

An old saw says that if the government really wanted to help literacy and reduce addiction in the inner cities, it would form a Department of Drugs and declare a War on Education. It's a fact that in random tests of airport security procedures after the tightening up that followed September 11, 2001, roughly half of the test weapons got through undetected.[7] Grand government prohibitions are primarily a security blanket and provide only a false sense of safety.[8]

Laws that do work, more or less, such as those against murder, work because almost everybody agrees with them, only a very few people try to break them, and those people are chased and mostly caught after they do break the law. Note the "after" part.

The more people want what is prohibited, the bigger the black market will be. Drugs are impossible to stop, but are seriously sought by a relative few. A better model for nanotech benefits might be Prohibition, when a major segment of the population was in the market. The laws were essentially laughed at. And yet some people believe that technological progress, and all its enormous benefits to virtually everybody, can be stopped by similar restrictions. These people are simply in denial.

With nanotechnology, the stakes are much higher. With a self-replicating threat, the enforcement web need miss only one development effort in the entire world to fail. And when that happens, the aerovores will be there, but we will have prevented any development of skysweepers. It's too great a risk to take. Indeed, it's not really a risk but a certainty: if open, aboveboard nanotechnology development is stopped or even seriously delayed, calamity is assured.

PAINT, SOAP, AND WASHING MACHINES

If, on the other hand, nanotechnology is developed evenly so that the mainstream, in general, has the same level of capability as rogue states and rogue groups, there's no great problem. Nanotech threats that would be catastrophic to a world equipped only with conventional bulk technology will instead be simply spills to be mopped up, fires to be put out, and generally preventable by ordinary hygiene and maintenance.

Imagine that it is 1900. Some people (like the Wright brothers) believe heavier-than-air flight is possible, and others don't. Some people have become alarmed by the idea, and they spin out a scenario in which a group of bandits build a plane, at a level of, say, a WWII bomber such as a B-29. In 1900 such a group could have leveled Washington without resistance and dictated terms to the country.

An alarmist at that date might well have drawn such a picture and urged that heavier-than-air research be banned. People actually building kites and wood-and-silk gliders would have scoffed at the proposed capabilities of a B-29, and the scoffing would jibe well with the common sense of the day. That alarmism and scoffing can clearly be seen in public discussions of nanotechnology today. Neither side has a handle on the truth.

By 1950 a rogue group with a single B-29 would have been blown out of the sky by jet fighters. Yet equivalent airplanes plied the air every day carrying passengers and freight. No one worried about some criminal group buying an airliner and using it as a bomber, because air forces were in place and the opportunity for major mischief was gone. Criminal groups had much more profitable things to spend their money on.

The notion that someone could have built a B-29 in 1900 and wielded it against an unsuspecting world is ludicrous. So are the scenarios you hear about full-fledged Stage V nanosystems loosed on today's world. Technology imagined out of temporal context sounds a lot scarier than it actually is. Let's take another example, this time not of evildoing, but of an accident.

Used tires are a major problem. They're hard to get rid of, they don't recycle easily, and they accumulate in giant piles that sometimes catch fire. (The fires are quite dangerous and difficult to put out.) Suppose some bright person decides to convert this problem into a resource. He invents a tire rot that lives on tire rubber and, using some of it for energy, converts the rest into oil. Then you take a big oil storage tank and start throwing tires into it, and first thing you know, you've gotten rid of the tires and have a full tank.

You know what happens next. Some of the rot escapes and finds its way onto a tire on a vehicle. The tire starts to rot away, leaving a trail of oil all full of rot nanobots on the road. By the time the first tire is gone, every road in the state has little oil tracks and anyone who drives there picks it up. In a week it's all over the country, and cars and trucks are worthless.

That's what happens if we assume nanotechnology completely out of its own time, or a situation where mainstream nanotechnology has been suppressed. There are several much more realistic scenarios, in a world where nanotech development is embraced by the mainstream:

1. The inventor, trained in engineering schools where safe design was taught, and using widely available tools, finds it easier, as well as the only legal way, to design a system with built-in limitations and safeguards.
2. The new rot is quickly detected by the ubiquitous environmental sensor net. It is analyzed, and poison molecules, carefully designed to affect only this one replicator, are released through the network. People in affected areas are warned, they download the recipe for the poison to their home synthesizers, and spray it on their tires.
3. Rubber rot is out there already, so tires are built with immune systems. At a decent level of nanotechnology, it's stupid and wasteful to build a tire as a balloon of bulk material anyway. Instead, they are nonwearing, self-repairing machines that literally grab the road at a microscopic level—a lot smoother riding and safer than balloon tires.
4. Everybody uses hovercraft or flying cars by now, so nobody cares. The last of the old rubber tires are eaten away, and the spots are sprayed with standard oil-slick-removal nanobots.

The more advanced the general nanotech level of society in general is, the less a problem the release—which would be a disaster today—would amount to.

Wood rots. If you take a piece of untreated wood and set it on the ground in a natural environment, it will be spongy and crumbly within a year. If you cut your finger and don't apply antiseptic, you'll soon find that microscopic machines—natural ones, bacteria—are waiting everywhere to munch on you, as well. New natural munchers evolve all the time. Simple poisons we've developed over the last century, antibiotics, go out of date as the organisms evolve around them.

The nanotech equivalent of paint for wood and soap for our bodies is clearly something we will be better off developing no matter what. By the time new mechanical munchers emerge, we'll be perfectly ready for them, if in the meantime we haven't been foolishly ignorant, or panicked with our heads in the sand.

> The Moving Finger writes; and, having writ,
> Moves on: nor all your Piety nor Wit
> Shall lure it back to cancel half a Line,
> Nor all your Tears wash out a Word of it.
> —Omar Khayyam, *The Rubaiyat*

CHAPTER 16

REAL DANGERS
The Same Old Bad Guys,
with Nastier Toys

It was a bright cold day in April, and the clocks were striking thirteen.
—George Orwell, *1984*

I t was a cold winter's night in New Jersey. It doesn't get bitterly subzero cold in New Jersey but it's always humid there, so the cold seeps into your bones. I had escaped it into the warm, brightly lit nostalgia of that New Jersey institution, the all-night diner. I had gotten a worried e-mail from a friend I will call Ed. He needed to talk about something that he wasn't willing to trust to e-mail.

It was the early 1990s. A few years before, I had formed the sci.nanotech discussion list on Usenet. Before the popularity of the World Wide Web, the Usenet discussion lists were the premier online forums for many technical subjects. Sci.nanotech had about twenty-five thousand readers worldwide and several hundred regular contributors. Ed, who worked at a nearby research laboratory, was one.

So we sat in a booth in a far corner of the diner and Ed described in hushed tones the revelation that had come to him that day. He had realized that nanotech could be used for isotopic separation.

To an extent, this was a legitimate concern. The heart of a nuclear fission device (bomb or reactor) is a quantity of fissionable material such as U235. U235 is an isotope of uranium that constitutes about 1% of natural uranium (most of the rest being U238, which is not fissionable). So even if you have a chunk of pure uranium, you need to separate the U235 out in order to have something useful for a bomb.

It's much harder to separate one isotope of an element from another than it is to separate two different elements or chemical compounds. Typically, different elements will have different chemical or physical properties. So you put your mixture—ore, for example—through some process that has a different effect on one than the other. A simple example is using a still to separate alcohol from water. Alchohol boils at a lower temperature than water, so if you heat the mixture to somewhere between the two temperatures, the alchohol boils off to be collected in your condenser, while the water stays behind.

But with isotopes, the properties are so similar that it's virtually impossible to distinguish atoms of one from the other. Uranium itself is a very heavy, fairly inert metal. Boiling it is out of the question. To separate U235 for the original bombs during World War II, the United States built the Oak Ridge plant, a city-sized processing facility that still took months to produce enough U235 for one bomb.

The bottom line is that isotopic separation is one of the major technical challenges that forms a serious barrier to entry into the nuclear weapons club. And that's a good thing.

Now it turns out that if you only have a few atoms of uranium, or anything else, it's not all that hard to separate them. The difference between U235 and U238 is about 1% in weight. Ionize your atoms and run them through a mass spectrometer, which is an instrument that fires them through a magnetic field. The field makes them curve. The lighter ones curve more easily, the heavy ones don't curve as much; so they arrive at different spots on the other side of the field and are thus separated. The mass spectrometer is a scientific instrument that is very useful in analyzing tiny samples of a substance, but it's of no use for the kind of quantities you'd need for a bomb.

With nanotechnology, you don't even need to fool with the tricks like ionization and so forth that bulk-technology mass spectrometers use. Essentially you just build a gadget that grabs an atom and weighs it. If its atomic mass is 235, it goes in bin A, and if it's 238, bin B. Gravity is of no use for weighing atoms, but a centrifuge would work fine.

Or more precisely, you build a million billion such gadgets, together with the machinery to distribute atoms to them and put the results back together. This composes a machine that sucks in 5,000 pounds of chemically pure uranium dust on one end, dumps 4,950 pounds of depleted uranium dust out the bottom, and produces a 50-pound sphere of U235 at the business end. Woops, better make that several pieces that would assemble to make a 50-pound sphere, and make sure you keep them apart.

So Ed was worried. Didn't nanotech mean that anybody could make a nuclear weapon? Wasn't this something we'd better keep secret? What could we do?

I hastened to reassure him. Yes, nanotechnology would make isotopic separation easier; but this had been thought of plenty of times before as a general idea. There would be no particular use in keeping it secret. But even so, nuclear proliferation wasn't something I was particularly worried about, I said. Compared to the other weapons that nanotechnology made possible, nuclear bombs were hard to build, hard to hide, hard to deliver, and of limited effect.

SELF-REPLICATING WEAPONS

> In a nanotech war, presumably, nothing appears to happen for two weeks. Then everyone on the losing side disappears.
> —seen on sci.nanotech in mid-1990s

There is a notion that nanotech weapons will be bacteria-like replicators that can live off the land of the enemy's territory and ultimately reduce it to dust or have any other devastating effect that the weapon-wielder desires. There are two points that need to be stressed about such a scenario:

First, the weapon scenario is much more likely than that of some kind of accidental replicator release. The difference is that commercial nanomachines will be made for high efficiency (and profitability), and thus be engineered to run fast, from high-energy foodstocks, be controllable, and produce useful outputs. Trying to live off the natural world violates all these desiderata. Accidentally building a nanomachine that would operate in the wild would be about as likely as accidentally building a car that sawed down trees, cut and split them into firewood, and stoked a boiler firebox with them.

It would be possible to design such machines on purpose, however, and weapons have always been the application where efficiency and other such considerations are last on the list. If someone designs a nanomachine capable of operating in the wild, it will almost certainly have been done as a weapon.

The second point, however, is that while it would have been possible to design army tanks that burned wood and thus could operate in the forest, no one ever did. In fact, actual weapons systems tend strongly in the opposite direction. The reason is that although efficiency is at the bottom of the list of design criteria for war machines, power and speed are at the top, and stopping to eat grass or whatever isn't the way to get them. Commercial nanomachines will manufacture more of themselves from high-purity, high-energy raw materials and so will military ones.

There just isn't the energy available for the taking in the natural world to power machines at the level our current-day machines operate, much less nanomachines. Wood-burning machines can't even compete with coal-burning ones, much less oil- or hydrogen-burning ones. Although it's too far out and too technically uncertain to mention in other connections, I wouldn't be surprised to see nuclear-powered military machines (radioisotope, not fission). Note that the military currently uses (fission) nuclear-powered ships, which are not cost-effective for commercial use.

Where today we have intelligent bombs and missiles, nanotechnology could give us smart bullets. Adding AI in nanocomputers to the mechanical capabilities would give us robot soldiers that could fold up to the size of a rat to creep through cover, unfold to the size of a human to manipulate objects or weapons, and fly at hundreds of miles an hour. They could be virtually invisible using phased-array optics skins. Their motions could be too fast for a human eye to track (this isn't hard at all—think of an ordinary electric fan).

To be prepared for war, a nation need only have a widespread nanomanufacturing base that could produce such robots, with matching vehicles and weapons, should the need arise. When push came to shove, the actual army could be produced in under an hour.

A good-sized patch of phased-array optics could focus light on a target. If the roofs of buildings were covered with these, they could act as solar collectors, radars and optical cameras, spotlights, and if tilted enough, streetlights. If necessary, and if used in a coordinated way from many buildings at once, they would function as laserlike antiaircraft artillery. In a post-9/11 world it is regrettably understandable why that might be desired. Private spacecraft, in particular, would be viewed with concern, and might be allowed only if there were some "instant-on" distributed air defense.

WAR

In the twentieth century, the cause of human disaster that clearly stood out above all the others was war. Technology clearly contributed to the deadliness. Will nanotechnology make war worse, or more likely?

I strongly doubt it. After all, a nanoweapon can't kill you any deader than a hydrogen bomb. What's more, since the development of nuclear-tipped ICBMs, a remarkable thing has happened. Nations have quit going to war against other nations that are armed with them. The author is of the opinion that this has a lot to do with the fact that such weapons put the political leaders of the attacking country in direct physical danger. As Ambrose Bierce quipped, the meaning of *rear* in American military matters is that part of the army closest to Congress.[1] Nanotech weapons will only exacerbate this phenomenon, making it easier to target the politicians. If so, war could suffer from a further decline in popularity.

Another phenomenon to consider is the industrial revolution and the Pax Britannica. If one nation or aligned group does attain a level of technological dominance comparable to Britain in the nineteenth century, a similarly peaceful period might ensue. At the moment, the United States and Europe account for a large majority of known nanotech efforts. Although certainly not perfect, the Western democracies would probably make a reasonably enlightened and stable world hegemony.

The one scenario that seems to merit some concern is the opposite—where other countries develop nanotech first. The United States might then engage in some serious saber-rattling rather than slip quietly behind. Here's a vignette that, thankfully, is completely imaginary at the moment:

It's 2015. In the United States, business as usual has been allowed to prevail. Interest in science has continued to decline. Virtually all the scientists and engineers our universities produce come from, and most return to, other countries. Funding for research is mostly for medical applications, and that is mired in political debates over stem cells and choked with red tape attempting to make it totally safe.

Meanwhile, China has pushed ahead on a broad range of fronts and has produced Stage III replicators. Products begin to appear from China that cannot be made economically anywhere else. No official notice is taken in the United States because our labs can still produce better stuff in expensive, one-off, form. The Chinese are accused of "dumping" and some nanotech products are banned.

China proceeds to Stage IV and Western technology begins to look distinctly second-rate. They are rumored to have engineering design supercomputers. The latest generation of Chinese jets and spacecraft has significantly better capabilities than ours and was designed and produced in half the time. The US military sounds an alarm.

The administration undertakes a crash program to demonize nanotech as "weapons of mass destruction" and get a UN resolution prohibiting it anywhere in the world. This stalls in debate and goes nowhere. The United States makes a unilateral ultimatum to China demanding a halt to all nanotechnology. China demurs, and announces that if attacked with nuclear weapons, it will release aerovores into the atmosphere.

It is perhaps worth noting that during the industrial revolution, France engaged in a series of wars—that were ultimately won by England at Waterloo.

TERRORISM

On September 11, 2001, I found myself in a limousine on the way to Newark airport. The driver's phone rang in the notes of Beethoven's "Moonlight Sonata"—it's amazing what details one remembers at times like these. The driver's wife, watching television at home, had called to tell us that a plane had flown into the World Trade Center. We looked up, and the New York skyline was in view. The World Trade Center looked like a pair of smokestacks. We got to the airport, and, of course, the flight was canceled. As we drove back, watching across the New Jersey marshlands and the Hudson River, we saw the towers collapse.

The question of terrorism can be divided into two parts: why do people try it, and how do they succeed? The first is a constant of human nature: take people and make them feel powerless, put upon, exploited, and having nothing to lose. Religion is often involved but any strong ideology will do. Some of them will explode in whatever way they can. It's the nature of the beast. The second is more complicated, depending on technology and other circumstances.

If instead of concentrated, regulated, commercial air travel, all the passengers flying that day had been in private aircars, September 11 would have amounted to a handful of carjackings, if that. Personal aircraft the size of cars simply could not have brought the towers down. Al Qaeda did not field enough operatives to destroy the towers on a piecemeal basis. You may remember they had already tried explosives-filled vans some years before.

That's not to say that private aircars would have been economically feasible in 2001, but to point out that decentralization is one of the best defenses against terrorism. A terrorist is at his most advantageous position when the defense has made a whole population helpless in hopes of disarming whatever terrorists may be among them. Public transportation is a favorite terrorist target: airliners in the United States, subways in Tokyo, buses in Jerusalem, trains in Madrid.

Oppression, or perceived oppression, generates the kind of anger, shared by a whole community, sufficient to marshal significant resources and motivate suicide agents. In the short run, this is purely a matter for statesmanship. In the long run, the kind of independence nanotechnology can provide may help. In the medium term, the problem of making the transition is delicate and difficult, but must be faced; because in the long term, some terrorists will get nanotech weapons anyway.

What happens next, we've already seen. If we stick our heads in the sand, we're in trouble. If nanotechnology is a widely based mainstream development, the effects are not likely to be worse than what we see today. If nanotech's decentralizing capabilities are used, terrorism could get harder to pull off. It would also be easier for the mainstream to retaliate.

As of this writing, after more than two years and after conquering two countries on the other side of the world, American forces still have not captured Osama bin Laden, despite a $50 million reward. Some of the more useful intelligence we have obtained from the badlands of Afghanistan has been through remotely piloted survellience drones: semirobotic aircraft. With nanotechnology, we could flood the country with drones the size of flies. Hiding would be much more difficult.

Current-day drones have been equipped with air-to-surface missiles and used to strike buildings and vehicles. Nanodrones could carry stings of various kinds and target individuals for lethal, debilitating, or punitively painful doses of chemical or biological agents.

A final note on terrorism: In my opinion, the major high-tech terrorism threat is not nanotech, but biotech. Not only haven't we caught bin Laden, we haven't caught the anthrax mailer, either. He or she seems likely to have been an American working alone. Alone!— while bin Laden and the Taliban had thousands of followers and supporters.

The nanotech version of a bacterium, a loose mechanical replicator of microscopic size, might be more efficient than a natural one. But there can be no nanotech version of a virus. A bacterium carries its own building machinery, which nanotech can make in a more efficient form. But the virus is even better—it carries no machinery at all. It uses the machinery in your own cells for its purposes. It consists of, in simple terms, just the genes needed to make copies of itself, and your cells do all the work. That's a level of finesse nanotech couldn't use. But note that viruses could be made right now with DNA you can order online—$2.35 per base pair—and with equipment in a high school bio lab.

As far as technological timescales are concerned, biotech is already here. It's as easy or easier to hide than nanotech. And people haven't come close to realizing the scope of threats it could produce. A properly tailored superflu could kill millions, given, say, twenty suicide agents to start the spread all across the country at the same time. Or someone could develop a mad cow prion for chickens. Or a Cipro-resistant anthrax. Or transplant the botulin toxin gene into algae and drop samples into reservoirs. Although these aren't nanotech threats, nanotechnology (nanomedicine in particular) would be invaluable to help fight them.

ECONOMIC STATUS

In the foregoing, I've addressed the applications of nanotechnology as if it were the United States and/or Europe that developed and deployed it. However, that's far from a certain outcome. Although we have a great head start at the moment, we could easily wind up as the Scotland of golf, the England of tennis, or the France of the industrial revolution. The reasons, as explained earlier, are manifold: bureaucratic business as usual here; a decline in the number of Americans going into science and technology juxtaposed with a huge increase in the number of science and technology students in Asia; and, probably, a greater willingness to take risks there.

Add to that, of course, the fact that Asia is not monolithic, but has numerous competing nations. This has several effects: competition is a strong motivator and acts as a spur to prevent bureaucracies falling into the lassitude which is their normal state. Look at what NASA achieved in the 1960s, when there was a "space race," compared to what it has done since. Second, in a project of research, one country might commit to one approach, and another, knowing it'd be behind if it just followed suit, could try something else that might turn out to work better. But the advantage is more subtle than simply trying different things. In a hierarchical organization, when a contrarian scheme begins to look like it might have been a better idea than the reigning orthodoxy, there's a strong temptation on the part of the higher-ups to quash it. (See the story of Semmelweis and the washing of doctor's hands, in "Nanomedicine," chapter 17.) In a situation where the competing elements are separate nations, this can be avoided to some extent.

If nanotechnology were developed in Asia with a significant lead time, the United States would be reduced to the status of a second-rate power. We are used to living behind the buffers of the great oceans and projecting our power where we wish on the globe. We could lose both—the dominant power might simply decide to tailor the Earth's climate to suit itself, for example, without bothering to ask us. You can be sure that the economic realities would shift, and we would be on the wrong end of some bad deals, shut out of monopolies, and, in general, have a lower standard of living and fewer opportunities than if we maintained the edge.

NANOPARTICLES

One danger that has been mentioned in the press in connection with nanotechnology is the release or biological effect of nanoparticulates. Today's nanoscale technology is producing some new products, such as paint that resists bacterial growth. But the effects of such substances, while different in detail from current substances we deal with, don't differ in any significant overall way. After all, ordinary untreated wood contains substances that inhibit bacterial growth. Ordinary chemicals are made of molecules even smaller than nanoparticles.

The big difference with real, eutactic, nanotechnology is in the opposite direction. Current technology cannot produce products and energy without releasing floods of molecular-level junk into the environment. If you want to produce billions of chemically reactive nanoparticles, just build a fire in your fireplace, or drive your car, or cook something. If you can smell it, it's releasing molecules that fill the air.

Eutactic processes change that. Every molecule is accounted for. Rather than dump odd chemicals and particles into the environment, they can be broken down to small constituents and used to build the next thing. Instead of producing carbon dioxide, for example, the nanofactory would release pure oxygen, and save the carbon to make diamond. The only combustion product normally released would be water.

OTHER DANGERS

> SWEEPING across the country [comes] the mechanical device to sing for us a song or play for us a piano, in substitute for human skill, intelligence, and soul. . . . I foresee a marked deterioration in American music and musical taste, an interruption in the musical development of the country, and a host of other injuries to music in its artistic manifestations, by virtue—or rather by vice—of the multiplication of the various music-reproducing machines.
> —John Philip Sousa, "The Menace of Mechanical Music,"
> *Appleton's Magazine*

Suppose you were transported back to 1900 and asked to explain to people what dangers were posed by the motorcar. If you said pollution, you'd be laughed at. You'll understand why if you've ever ridden on a train with a coal-burning steam locomotive. Steam engines of the day produced a lot more noxious smoke than gasoline engines. A report of deadly crashes might find more favor—people were scared of these unusual mechanical monstrosities. But you'd still be wrong: cars have saved a lot more lives than they've taken, and the average American's lifespan has increased by more than twenty years since 1900. Traffic jams? Hard to imagine in a country with thirty acres per person. Fragmentation of the traditional extended family and dissolution of sexual mores? Excuse me, sir, are we speaking the same language?

The fact is that if you had to wait to die in a car crash, you'd live an average of 6,700 years. Traffic jams are a nuisance, but except in the most densely populated areas, there's no great demand for alternatives such as public transportation. In my own experience the greatest problem with cars in cities is parking.

The social effects are perhaps somewhat closer to the mark. In fact, the twentieth century has seen changes in our lifestyles that would have horrified our great-grandfathers. But are we the worse for them? In some cases, perhaps; in others, probably not. In most cases there are compensations—the changes have made things different rather than particularly better or worse. What's more, the changes came about from a confluence of causes, including cars, two world wars, radio and TV, the pill, the jet airliner, the computer, and the Internet.

The printing press has been blamed for the Reformation and some hundreds of years of religious wars.[2] This could obviously have been handled better, but these were the growing pains for moving from a largely illiterate culture to a literate one. Should we have passed up on it? I think not. It does give us a clue as to the real dangers that lie ahead, though. Entrenched power will react strongly to change, even if the change would ultimately benefit most people.

Imagine the politician, or corporate or labor leader, who became possessed of a device that would allow him to put you in a chair, clap a helmet over your head, and five minutes later have you step out, a fanatical devotee of his, anxious to grab others and hold them down in the same chair. This is not something that anyone has any idea how to build right now, even given full-fledged nanotechnology. But biological knowledge is advancing as fast as any other realm of science, and faster than most. I'd find it surprising if by the end of the century, something like this weren't possible.

Suppose a great nation decides to do the same with its army, promising to set them back to normal after the "emergency" is over.

Suppose every citizen is put through a milder form of reprogramming "to prevent terrorism."

Nanodrones, described above as an antiterrorism weapon, could easily be misused as an assassination weapon. Nanodrones are early, not advanced nanotech, probably Stage III or thereabouts. Crude, only by comparison, microdrones can be and are built today. These have cameras only, as far as the author is aware. Even they threaten significant social change owing to ubiquitous surveillance.

Being constantly spied on is one thing. Being subject to termination with extreme prejudice on someone's whim is another. Nanodrones could be quite difficult to stop, and virtually impossible to trace. If they became generally available, it might be a bad time to be a lawyer. More seriously, note that the United States already has used its (full-sized) drones for assassination, and that governments as a class, over the twentieth century, killed tens of millions of their own citizens and at least as many of the citizens of other countries. Even if governments can be forced to open their operations to public scrutiny and oversight, other organizations will make obtaining nanodrones a high priority. It will be a bad time to be Jimmy Hoffa.

Note that like the other threats, nanodrones are not nearly so formidable to a population that is equipped with nanotechnology. A drone would be useless against someone living in a nanoskin or surrounded by Utility Fog, for example. The danger is that governments might, in a protectionist panic, limit access to countermeasures. This could easily be a slippery slope to a situation with nanotech-equipped officials and powerless, unprotected commoners. And government of the people, by the people, for the people would perish from the Earth.

LIVING IN INTERESTING TIMES

> Nearly all men can stand adversity, but if you want to test a man's character, give him power.
> —Abraham Lincoln, quoting François Duc de la Foucauld

The real dangers that will come with nanotechnology are not the kind of things that play well on a movie screen or can be explained in sound bites. They are not dangers from the technology itself, but the effects of shortsightness and greed in the face of a revolution in human affairs.

Every major, life-changing development has been accompanied by some strife. Much of the strife has been, in retrospect, unnecessary; but it never seemed that way at the time. In many cases, as in the Reformation, people didn't really understand what they were fighting over until long after the dust had settled.

One of the best ways to prevent, or at least minimize, strife as nanotechnology is developed is for there to be a broad understanding of what benefits it can bring and what the dangers really are. If nanotech remains the dimly understood magic of a powerful few, trouble lies ahead. But if people widely understand how personal manufacturing technology can bring independence and a comfortable lifestyle to everyone, and the way is made clear for this to happen, cooperation—and sanity—may just prevail.

CHAPTER 17

NANOMEDICINE
Fixing What's Wrong

I t was 1847, less than two lifetimes ago. In the General Hospital of Vienna, one of the leading centers of science and medicine of the day, there were two maternity wards. Mothers in First Ward were attended by doctors and medical students. Mothers in Second Ward had to make do with midwives. Yet the women of Vienna begged to be admitted to Second Ward, since First Ward had a reputation as a deathtrap.

The reputation was true. More than 12% of mothers in First Ward died of puerperal fever after childbirth, while less than 3% of those in Second Ward did.

A brilliant young Hungarian doctor, Ignaz Semmelweis, had recently been put in charge of the maternity wards and was desperate to find the cause of the deaths. He wrote:

> Almost every day, the sound of bells intimated that the priest was administering the last sacrament to the dying. I myself was terror-stricken when I heard the sound of bells at my door. . . . This worked on me as a fresh incentive that I should, to the best of my ability, endeavor to discover the mysterious agent, and a conviction grew day by day that the prevailing fatality in First Ward could in no wise be accounted for by the hitherto adopted etiology of puerperal fever.[1]

Semmelweis tried everything he could think of. The prevailing medical theories involved things like "miasma" (bad air) and imbalance of "humors" (bodily fluids). Miasma didn't work: the air in Second Ward was stuffier than in First. It was more crowded, given the women's quite justified preference. At one point he even hushed up the priests on the theory that the sound of the bells might be contributing. Finally he got a clue when a colleague, Jacob Kolletschka, died from a fever after cutting himself during an autopsy. Semmelweis noted that Kolletschka's symptoms resembled those of the dying women. He guessed that some infectious agent was being carried on the doctors' hands from the autopsies and educational dissections into the maternity ward.

Semmelweis instituted a policy that doctors and medical students wash their hands in chlorinated water before examinations and deliveries. The death rate plummeted to below 2%. The head of the hospital, Dr. Johann Klein, was stung by Semmelweis's insistence that the doctors had been causing the deaths and refused to believe his new theory. The feud became bitter, and the politically connected Klein ultimately forced Semmelweis from Vienna. Semmelweis returned to his native Budapest, where he joined St. Rochus Hospital and repeated his dramatic success against puerperal fever.

In the 1860s a similar controversy raged in the pages of the British medical journal the *Lancet*, as one Dr. Joseph Lister attempted to persuade the medical profession to use carbolic acid as a disinfectant in the treatment of open wounds.[2]

As late as 1879, it was still possible for a distinguished doctor to address the Paris Academy of Medicine and attempt to heap scorn on the idea that infection could be carried on doctors' hands from one patient to another. On this particular occasion, however, the doctor was interrupted from the audience by another who sprang to the podium and drew pictures of streptococci, as seen under the microscope, on the blackboard. The interrupter's name was Louis Pasteur.

By the turn of the twentieth century, the germ theory of disease and infection had been largely accepted, and there was a revolution in surgery. Along with antisepsis had come anesthesia. A patient could now enter surgery with a reasonable expectation of surviving it.

By 1950 another revolution had occurred. Sulfonamides,[3] penicillin,[4] and streptomycin[5] had been discovered and put to widespread use. In 1890 the leading cause of death had been tuberculosis, with 245 people out of each 100,000 dying of it each year. By 1990 the corresponding figure was less than one.[6]

A MOLECULAR WAR

In 1928, when Alexander Fleming discovered penicillin, neither he nor anyone else understood how it worked. Not until later, when the molecular machines that make cells work were being studied, was a full understanding reached.

What happens is this. Bacteria have a tough cell wall, essentially a skin, that is constructed of large, complex, interconnected molecules. To build the wall, the bacterium creates long molecules of saccharide subunits, hangs an amino acid subunit off each one making a comb-like shape, and knits the combs together into a netlike structure called peptidoglycan. The molecular machine that performs this last step is called transpeptidase. The penicillin molecule blocks transpeptidase by getting wedged into its active site, preventing it from working. You can think of this as similar to what happens to a stapler when you try to staple a piece of taffy. When there's enough penicillin around, all the bacterium's transpeptidase "staplers" get jammed, and it can't join up its peptidoglycan nets to make its skin. So when it tries to divide, a process requiring lots of new skin, it ruptures.

We can flood our bodies with penicillin because the cells of higher animals don't use peptidoglycan, and thus don't have any transpeptidase to block. Bacteria, on the other hand, can become resistant to penicillin by evolving an enzyme that breaks it down. Then we can attack them by adding another drug that inhibits the new enzyme, and so back and forth.

This is how most modern drugs work: by being molecules of just the right shape to form a monkey wrench in some specific molecular machine. We are lucky that there are enough differences in the workings of humans and bacteria to allow us to bollix them up without doing the same to ourselves. Today's medical research studies the molecular machines and reaction pathways that form the workings of cells, and invents new molecule shapes specifically for a particular effect instead of just trying random chemicals to see what works.

Drugs have a couple of major drawbacks, however. They are good at blocking molecular machines; it's harder to fix one that isn't working right. This is fine as long as the disease is caused by some bacterium or other form of life that the drug acts as a selective poison for. But having cured many of those, we are left with the diseases that occur when our own machinery goes awry. In 1890 heart disease and cancer killed 170 people per 100,000 per year. In 1990 the figure was 514.[7]

Mind you, the people who die of heart disease and cancer today are dying much later in life, on the average, than the similar number who died of tuberculosis and pneumonia a century ago. But the basic difference is that it is a lot easier to kill it than to fix it, especially when the thing to be fixed, the human cell, is literally a thousand times as complicated as the one, the bacterium, to be killed.

There are, of course, some "fix it" or "maintain it" things that can be done in a druglike way. Vitamins, minerals, and similar nutrients can provide small molecules that the cell can't manufacture or isn't manufacturing enough of. They can provide odd elements, like the iron in hemoglobin, in a form the metabolic machinery can access. Hormones, small druglike molecules that turn on and off various molecular-level functions, can be provided. Insulin is an example. (Drugs that block hormone activity by jamming their receptors are also common.) But you can't take fully formed biological molecular machines as medicine: your digestive system breaks them down just like any other protein you eat.

SURGERY

The wing of medicine that does "fix it" nowadays is surgery. Surgery is quite amazing in its capabilities: reconnecting severed limbs and transplanting organs are not uncommon. For some applications, notably the joints, arthroscopic surgery can be done. The arthroscope is a pencil-sized device that acts as lights and camera for the surgeon to operate through small incisions. This can enormously speed healing times as compared with techniques with large incisions and direct observation.

Nanotechnology will advance the arthroscope in obvious ways. It should be possible to do some kinds of surgery by injecting a thread no wider than a hair. The thread would be a conduit for a vast army of nanomachinery that would build it out into a network around the region of interest and proceed to reconstruct the diseased or damaged tissue. By and large, with this technique, the nanomachines would not add and remove material as much as is typically done in current surgery, but rebuild new tissue using the material of the old.

Other surgery would benefit as well, or perhaps even more. In particular, current surgical techniques disrupt cells, sever extracellular connective tissue, cut nerves and capillaries, and in general, do all sorts of damage that needs to be fixed by the healing process.

The surgeon with nanotech tools will not cut into you with a knife blade that is a slab of steel. He will use a machine, smaller than a current scalpel, that is more complex than an aircraft carrier. It will cut around cells rather than through them. It will send a detailed analysis of the tissue it is working on back to a database bigger than today's entire Internet. It will be able to terminate nerves and capillaries temporarily and flag them. When the operation is complete, other machinery will be able to reverse the separation, using the saved information and implanted flags, to a state that is nominal for healthy tissue. In other words, little or no healing will be necessary because the surgeon will put you back together, fixed and working, at the cellular level. There is no reason in principle that you couldn't have major surgery one day and play tennis, go dancing, or do a full day's work the next.

The same technology that will be used to implement the ultimate virtual reality, namely, tapping and spoofing sensory and motor nerve signals, would be invaluable as a replacement for anesthesia and its attendant dangers and recovery times. Not only would this speed the process and make it more comfortable, but you could be doing anything you wanted, via virtual reality (VR) link, while it was happening. Suppose you'd rather be playing tennis than lying on the operating table. A robot your size and shape walks onto the court, and it has a data link to the anesthesia equipment. Signals from the robot are injected into your sensory nerves, so you see and hear what it does. Likewise, all the signals from your brain to your muscles are diverted and transmitted to the robot, so it actually does the actions you intend. The real sensory signals from you body, those of being operated on, are jammed, muted, or disconnected, and those from your brain similarly squelched, so that your body doesn't try a diving volley down the line just as the doctor is reconnecting your pancreas.

SPARE PARTS

Today, there is in some cases a choice between artificial and donated organs. With nanotechnology, virtually any organ will available in an artificial form, in general with better performance than the original.

Organs that perform physical functions, such as bones and muscles, are the easiest to replace. These are the ones today's technology handles best, like joints. Applications requiring power, like pumps such as an artificial heart, are complicated by methods of providing the power. This is ironic, since the cells of the body are crammed with mitochondria that turn sugar and fat into electricity—there are literally billions of highly efficient fuel cells at work inches from the prosthesis, but current medical technology can't take advantage of them.

Just like the rest of our technology, medicine is hampered by the gap between mechanical and molecular capabilities. We're much better at breaking than fixing at the molecular scale. We could probably create a drug that would jam the ATP synthase motors in our mitochondria (it would be a deadly poison, of course), but we can't use them to power our artificial hearts. Nanotechnology will close the gap. Nanotech organs could be powered by just the same fat- and sugar-burning reactions that cells use. Powered organs like heart and muscle will be straightforward. Organs that perform chemical functions, like the liver, and those that do physical sorting at the molecular level, like the kidneys and lungs, will be next.

Replacing organs that are bodywide and connected to everything else, like the blood vessels, nervous system, or skin, will be the hardest of all. The problem isn't so much the ability to make such an organ as the physical capability to replace it. Imagine you had to repair a faded tapestry by snipping out the faded thread, one strand at a time, and replacing it with new thread. You have your sewing scissors, plenty of thread, and a needle. Ready? The tapestry you have to fix covers the whole Empire State Building.

The same problem happens when we contemplate doing something to the body at the micron scale (much less the molecular scale!). It simply isn't humanly possible to do that amount of work. The technology exists today to take a single cell and add or remove individual organelles under a high-powered microscope. If you could do that in a minute, working twenty-four hours a day, it would take you one hundred million years to do every cell in a human body.

Thus, once we figure out just what it is we want to do to cells, it's going to have to be done robotically. There just isn't any chance of doing anything requiring human attention to any significant fraction of the cells, or even capillaries, in a body.

SINGLE-CELL ORGAN REPLACEMENT

Red blood cells carry oxygen by absorbing it into hemoglobin, a complex protein constructed to do so. However, we could carry far more oxygen in the same volume by pumping it into a pressure tank. With nanotechnology, we can make the oxygen filter, pump, tank, some sensors, and control machinery, all smaller than a red blood cell. (Smaller because red cells fold to go through the smallest capillaries, but the tank wouldn't.) A mechanism of this kind is called a respirocyte, invented by Robert Freitas.[8] A respirocyte is essentially an artificial red blood cell.

The respirocyte might be about one cubic micron in volume, rounded smoothly to pass easily through capillaries, and shaped like any cylindrical pressure tank. The inside would have a partition that slides from one end to the other, separating volumes which contain oxygen and CO_2.[9] It would have pumps and filters for oxygen, CO_2, and glucose, which it would use as a fuel. In terms of our five-million scale example, it's the size and complexity of a propane delivery truck.

The respirocyte would detect when it's in the lungs (by the high oxygen level) and pump in the oxygen and out the CO_2; it would do the opposite in the rest of the body. A respirocyte's tank might have a capacity of half a cubic micron, holding oxygen at 2,000 atmospheres pressure. (If the tank leaked, the resulting bubble would be just 10 by 10 by 10 microns, the size of a single ordinary cell, easily dissolved into the plasma and reabsorbed by other nearby respirocytes.) Replacing just a quarter of your red cells with respirocytes would give you storage for about 3 pounds of oxygen right there in your bloodstream. For comparison, the body uses about a pound of oxygen a day at complete rest; or consumes it ten to twenty times as fast at the peak of exercise.

Besides letting you hold your breath all day, respirocytes would be good insurance against heart attacks. The oxygen in storage would sustain you for long enough to get help. The same is true if you were shot in the heart.

Another Freitas invention, the microbivore, replaces or augments white blood cells and antibodies.[10] It's essentially a nano-Battlebot that physically destroys invading bacteria, viruses, and fungi. A shot of microbivores programmed to recognize a specific pathogen could cure a disease like measles or flu in minutes. People who live in the

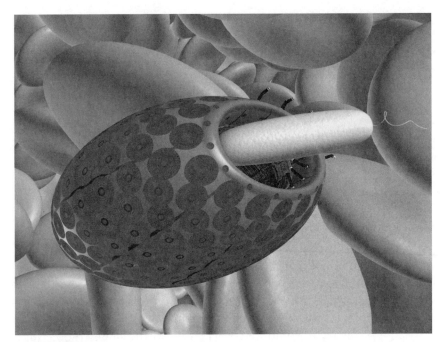

Figure 25. A microbovore. A defensive nanobot patrols the bloodstream, destroying harmful bacteria. Design by Robert Freitas, courtesy of Forrest Bishop (http://www.iase.cc/).

Northeast might get prophylactic doses for the Lyme disease spirochete, people in the tropics likewise for malaria.

Many other cells in the body could be replaced on a one-for-one basis. Some do simple functions such as pulling when told, that is, muscle cells. It would be perfectly feasible, for example, to supplement muscle with artificial winch cells. The trick would be putting them in place. For this purpose, replacement cells would have to be able to move within the tissues, find the right place to be, settle into place, and connect themselves to whatever was necessary. They could use a combination of internal recognition and external signals to locate themselves. If you did that, of course, you'd almost certainly want construction nanorobots to strengthen your bones and other ones to accelerate the maintenance on your joints.

Remember just how much machinery nanotechnology could put into the volume of an average human cell, a 10-micron cube. Scaled

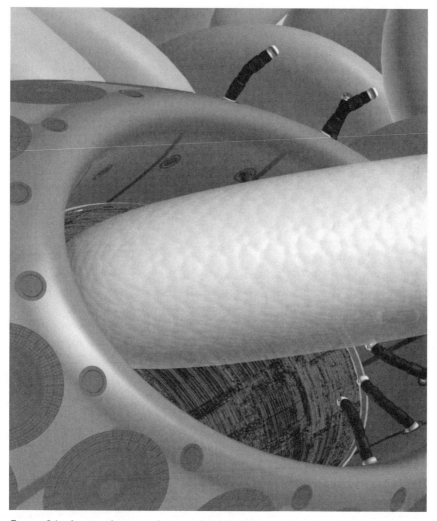

Figure 26. A microbovore close-up. In this close-up, the telescopic grapples can be seen engaging the bacterium. Design by Robert Freitas, courtesy of Forrest Bishop (http://www.iase.cc/).

to the size where its parts are like car engine parts, that's equivalent to a building seventeen stories high covering most of an acre. Or think of half a million car engine compartments. That much mechanism gives you the ability to do some very complex functions.

Note that the body contains signals—chemical and otherwise—

for a cell to be able to tell what it should be doing. One of the most promising directions in current medical research involves stem cells. These are cells, like those in the fetus, that have not differentiated yet and thus still have the potential of becoming muscle, nerve, bone, fat, or whatever. Stem cells can regrow damaged organs.[11] (It's the fetus connection, of course, that makes them controversial. However, the adult body also contains some stem cells, and there seem to be ways to convert other cells back into stem cells.) The point is that, although it's not completely understood yet, there's obviously enough information for the stem cell to "know" how to develop in its regrowth.

CELL REPAIR MACHINES

Ultimately, to attack things like cancer and aging, we are going to have to build robots that go into individual cells and fix genetic errors and repair other kinds of damage, such as those caused by free radicals and radiation. This is a bit trickier than replacing a cell. The size of a nanomachine that could move around inside a cell and do things is on the scale of one micron. This leaves it with five hundred car engines' worth of mechanism instead of half a million. It will obviously be possible to build machines that do particular functions, but a general-purpose repair robot that "understands" everything in the cell, and constantly checks and fixes it, is unlikely.

On the other hand, mechanisms that check for certain specific conditions seem quite feasible. One straightforward way of fighting cancer would be to place diagnostic units that detect when the cell's regulatory mechanisms have failed, and kill the cell (or block its reproduction). Cells already have such mechanisms. Half of the current incidence of cancer is associated with a failure of the "p53 tumor suppressor" system, for example.[12] A nanomachine could either duplicate (or extend) the p53 system's function or check the p53-generating genes for mutations and correct them.

The p53 isn't the only anticancer system in the cell—generally more than one thing has to break down before cancer develops. There are many things to check and fix, or to provide backups for. In the last-ditch case, nanomechanisms could be programmed to find cells exhibiting known molecular markers for cancer and destroy them.

Cancer isn't the only untoward effect of nasty stray molecules, free radicals, and radiation. The molecular-scale damage these agents cause has various other bad effects as well. Again, these things are normally fixed by existing mechanisms in the cell or by letting a cell that's too badly messed up die, and others grow in its place.

In time, though, damage accumulates in spite of repair mechanisms. What's more, some of the cells' defenses play us false. For example, the Hayflick limit is a sort of counter that lets cells divide only so many times. It's one of the defenses against cancer. But after the human "design lifetime" of seventy years, the Hayflick limit begins to get in the way of the replacement of damaged cells with new ones.

Aubrey de Grey, a leading gerontologist, lists seven major problems at the cellular level that lead to senescence. They are:

1. Cells dying off (a normal process) and not being replaced.
2. Fat cells replacing working cells, and cells that have quit working but won't die on their own.
3. Mutations in the genes of a cell (its DNA) that make it quit working, produce things injurious to other cells, or become cancerous.
4. Mutations in the DNA of mitochondria. Mitochondrial DNA is much less well protected than the cell's main DNA, so it's more vulnerable. When this happens, the mitochondria don't produce the ATP the cell needs.
5. Garbage buildup inside cells. Cells constantly recycle the things they make inside, but every so often something gets made so wrongly that the normal mechanisms can't disassemble it. Over the years, this junk builds up inside a cell and keeps it from working.
6. Garbage buildup outside cells. Again, the body has mechanisms for handling stuff like this, but there are a couple of cases where the buildup can defeat the body's cleanup crew. The main ones are the formations of plaques in the arteries causing atherosclerosis and the ones in the brain causing Alzheimer's disease.
7. Long-term chemical damage to the proteins outside the cell that have various physical functions, such as connective tissue, artery walls, the lens in the eye, and so forth.[13]

These are arguably it. In the 1960s and 1970s gerontology regularly added new kinds of damage to the list as more of the mechanisms of aging were discovered. But the list has been stable for over twenty years. There's a reasonably good chance that once we learn to fix the damage these categories represent, we'll be able to extend not only our lifetimes, but youthful vigor, to a significant extent.

It's important to point out that all these problems appear to be susceptible to some kind of treatment using nonnanotech techniques: stem cells for number 1, targeted cell-killing drugs for 2, a whole raft of research aimed at cancer for 3, and so forth. These techniques are mostly still in the labs, but it's a very exciting time in gerontology research.

Like many of the other nanotech applications I've discussed here, an assault on aging is one that is right on the edge of current or near-future capabilities without nanotechnology. Thus we can have fairly strong confidence that we can do it with nanotechnology. Being able to measure, manipulate, and monitor at the molecular level can only improve our knowledge and capabilities.

Here's how we might expect nanotechnology to help address de Grey's seven pathologies:

1. *Cells dying off:* Stem cells are still the ticket here. However, a nanomachine may be able to prod a cell to divide where it would not have, by lengthening its telomeres, for example.
2. *Cells that have quit or turn to fat:* Finding cells that aren't working, or working well enough, is the hard part. "Census-taking" nanorobots that wander around checking cells might work. Killing the bad cells would be simple. There may even be some internal "switches" that can be thrown to change a fat cell back into muscle or whatever, or into a stem cell.
3. *Mutations in the genes:* Nanomachine to check or extend the cell's p53 system and other subsystems that repair DNA (or kill the cell if it's too damaged).
4. *Mutations in mitochondrial DNA:* Several options exist: check and fix mitochondrial DNA, manufacture the proteins it's supposed to specify, or simply replace the mitochondria with mechanical ones.
5. *Garbage buildup inside cells:* Cell repair machines would have two basic options: break the junk down in place or carry it away. Which is best remains to be seen.

6. *Garbage buildup outside cells:* You don't even need cell repair machines for this, just larger ones that physically clean up the accumulations.

7. *Damage to extracellular structures:* Larger construction and repair nanorobots, as in point 6, but the machines check the material for damage and patch as needed.

Freitas has gone further than merely halting aging with the idea of "dechronification," that is, rolling back the clock. Dechronification will first arrest biological aging, then reduce your biological age by performing three kinds of procedures on each one of the tissue cells in your body:

First, a cell maintenance machine will be used on each cell to remove accumulating metabolic toxins and undegradable material. Afterward, these toxins will continue to reaccumulate slowly as they have all your life, so you'll probably need a whole-body cleanout to prevent further aging, maybe once a year.

Second, chromosome replacement therapy[14] can be used to correct accumulated genetic damage and mutations in every one of your cells. This might also be repeated annually, or less often.

Third, persistent cellular structural damage that the cell cannot repair by itself such as enlarged or disabled mitochondria can be reversed as required, on a cell-by-cell basis, using cellular repair devices.

The net effect of such interventions could be the continuing arrest of all biological aging, along with the reduction of current biological age to whatever new biological age is deemed desirable by the patient. These interventions may become commonplace several decades from today. Notes Freitas: "If you're physiologically old and don't want to be, then for you, oldness and aging are a disease, and you deserve to be cured. Using annual checkups and cleanouts, and some occasional major repairs, your biological age could be restored once a year to the more or less constant physiological age that you select. I see little reason not to go for optimal youth—a rollback to the robust physiology of your early twenties would be easy to maintain and much more fun. That would push your Expected Age at Death up to around 700–900 calendar years. You might still eventually die of accidental causes, but you'll live ten times longer than you do now."[15]

ETHICAL ISSUES

There seem to be two kinds of ethical issues facing medicine today with its new capabilities. The first is questions of whether we should actually do some cure or improvement, for its own sake. The second is the question of making humans, or parts of them, specifically to cure other people.

I will leave questions of the first kind for the chapter on transhumanism, although as far as cures are concerned, the issues seem at least somewhat more straightforward. For the second kind, though, nanotechnology may be able to finesse some of the thornier problems.

The reason is that the nanotech route to certain kinds of capability involves building machines that are clearly machines and have specific, planned functions. The biotechnology route, however, leaves you in the gray area between handling parts and handling people. Questions like the morality of cloning, or buying and selling organs, tend to arise. With nanotechnology, where you can fix the patient's own cells and build purely mechanical parts as good as the originals, those particular questions don't stand in the way of healing.

LIVE LONG AND PROSPER

An American man born in 1850 had a life expectancy of about thirty-eight years. An American man born in 1990 has a life expectancy of about seventy-three.[16] In other words, for every four years that passed during the period, life expectancy increased a year. (It wasn't steady, of course—wars and medical advances like penicillin varied the rate.) Besides medicine, much of the improvement was due to the effects of the first industrial revolution: proper food, clothing, shelter, sanitary facilities, and so forth.

The demographic human lifespan is beginning to bump up against old age, as productivity and medicine have removed the bulk of the causes that used to cut it short. But we have seen aging is not mysterious; it is caused by understandable phenomena that we will soon have the means to combat. There's no reason, then, to expect the historical trend to stop. More specifically, if you can hang around for the next few decades, you can probably expect to be here for quite a while longer than that.

In 1960 total private and public spending on health was $27 billion. In 2000 the figure was $1.3 trillion, nearly fifty times as much. By 2010 it's expected to double again to $2.6 trillion.[17] The obvious question to anyone faced with today's high health insurance costs is whether nanotechnology will be able to apply the same kind of drastic cost reductions to medicine that we have projected for manufacturing and other high-tech applications. In the long run, quite likely. In the short run, not so likely. There are many structural reasons for this, but the basic cause will be that every advance nanotechnology brings will be swallowed up in the great struggle to understand and then defeat cancer, AIDS, and old age—not to mention heart disease, arthritis, and the common cold. It may be late in the twenty-first century, and technology enormously advanced on other fronts with capabilities we can now only dimly dream of, before these are conquered. After that, the cost reductions will begin to take hold and finally settle down to reasonable maintenance costs for the human machine.

After that we'll face the *real* problem: what to give on a 250th or 500th anniversary.

CHAPTER 18

IMPROVEMENTS
A Perspective on Transhumanism

MIRANDA: O, wonder!
How many goodly creatures are there here!
How beauteous mankind is! O brave new world
That has such people in't!

        —William Shakespeare, *The Tempest*

There is a broad and slippery slope between merely patching up someone who has been injured, or who suffers a genetic deficiency, and augmenting people to better capabilities. Humans vary widely in native ability and strength. Some are virtually immune to sunburn, others are debilitated after an hour outside on a bright day. Most are in between. The same is true of attributes ranging from height to intelligence. It's a cure to add respirocytes to someone's blood who's anemic. It's an augmentation to add them to a normal person. What about someone who's in the normal range, but somewhat lower than average? If you were to try to live at high altitudes, in the Andes or Himalayas, for example, you'd find that a good helping of respirocytes would just bring you up to the normal, evolved performance of your neighbors.

What about intelligence? Nanotechnology may well be playing second fiddle to biotech for true, organic brain upgrades for a while. Even so, people have been using technology to enhance their intelligence for millennia. It's amazing what just pencil and paper can do. You'll find it laborious to multiply 629,904,623 by 984,571,094 with pencil and paper; but you'll find it impossible to do it without them.

Of course, you'd never try nowadays. You'd use a calculator or a computer. More to the point, you'd rarely have to do multiplication at all. Calculations would be made by software that interacted with you on a much higher level, and you'd never know how many multiplications were being done on your behalf. Software can extend your intellectual abilities in many ways; consider a tax-form preparation program or an engineering CAD system.

Imagine you have a computer, a sort of super-PDA, that sits in your pocket. It has the storage to hold a large library of reference material, as well as wireless access to the Internet when needed. It has the latest in voice and image processing software, and the computational horsepower to run it in real time. You wear a pair of glasses that can overlay a generated image on the real world, giving you captions, briefings, memory jogs, whatever, without your having to seek out a screen or lug a laptop.

The system listens to everything you say or that is said in your presence. It watches what you see with tiny cameras in the glasses. It transcribes everything so that you can review conversations at leisure. It remembers everyone who you are introduced to and can place a virtual nametag on them when you meet them again. Anytime a question is asked, it looks up or calculates the answer and puts it in a subtitle in your field of view. It translates from other languages, spoken and written, the same way. Or indeed, reading in your own language, if your gaze lingers on a word and you mumble "huh?" the definition would pop into view.

A system like this is possible right now, with two exceptions: the computer to handle the job would be backpack-sized instead of pocket-sized and the software isn't really there, though all the basic capabilities exist at least in labs somewhere. With nanotechnology, you could put the entire system into a single contact lens. The smarter the computer gets, the smarter you act.

The next step, of course, is to put the computer inside. You can put it down at the brain stem, where it can listen in on all the nerve signals coming into the brain, using your eyes instead of cameras and your ears instead of microphones. Indeed, it can tap into the brain's cross talk at higher levels, taking advantage of some of the preprocessing your visual system does, for example. Then it can inject the hints and answers into the appropriate sensory nerves.

No need to rely on the biological human senses, of course. We can extend the range of vision, from telescopic to microscopic, and into the infrared or the ultraviolet. We needn't modify the eye; any patch of skin could be embedded with a grid of microscopic sensors and act as a phased-array optical antenna. (And look exactly like ordinary skin to normal human vision). You could know the exact chemical composition of anything you tasted or smelled.

It would be very useful to be able to look inside yourself and check out how all your organs were doing, from a surgeon's eye view to a microscopic examination of individual cells. In the opposite direction, there should be ways, mediated by information networks, of seeing the larger things you're a part of: community, company, nation, with an integrated and instinctively understandable display.

Enough has been written about virtual reality that it's probably not useful to say much more here, except to note that it would be a simple adjunct to the capabilities of this level of augmentation to inject a completely synthetic set of signals into your sensory nerves and transport you to an imaginary world, or simply one at a different place. There should be no more need for offices at all; if your stock in trade is information, you just take a walk in the park. Anything you could have done on a computer could be seen in the mind's eye; for your meeting, sit on a rock in a forest glade and your conferees appear in virtual form around you.

Just as useful, and probably a lot of fun, too, would be imaginary friends. With nanotechnology you can have enormous computer power stuffed into odd spots in your body, powered by all those desserts you don't have time to exercise off. This can be used to simulate a number of companions, who can work on intellectual stuff while you're doing something else. You can "hear" and "talk" to them subvocally, and they can give you presentations in your field of view. The closer they get to real AI, the closer you are to being a large, well-integrated team instead of just one person when you tackle a problem.

ELBOW GREASE

Current medical prostheses are mostly physical: arms, legs, and so forth. More sophisticated ones are appearing, starting with cardiac pacemakers and ranging to cochlear implants, as the technology improves to be able to simulate, to some useful level, the more complex functions of the body.

With full Stage V nanotechnology, virtually the entire body could be replaced and have its capabilities enhanced. The reason to do so would not primarily be to extend strength or endurance—we would have squads of robots to accomplish whatever we needed physically—but survivability and recuperation in the case of injury.

The major watershed would be whether to retain the ability to look and feel just like a standard biological body. Given the penchant for body modification among the upcoming generations, this may not be as important a shibboleth as it is to us older folks. One physical ability I would be willing to undertake drastic modifications for would be to be able to fly. Biological human bodies, even with the most extreme imaginable biotech modifications, are very unlikely to be able to handle it. Human flight is a nanotech-only proposition.

A much closer and more likely motive is aging. Biotech has a number of approaches that seem to be promising, but in the long run, it may just be simpler to replace the evolved kludge with a new model that doesn't have the biological clocks that need resetting all the time. You can be immune not just to aging but to all known human diseases, eat anything (including wood, coal, candle wax, or gasoline), and live comfortably anywhere on the surface of any planet in the solar system, or in empty space.

All of this is simple and easy to take, however, compared to replacing the brain. Again we face a slippery slope. The modifications discussed above as the "super implanted PDA" appeared to enhance intelligence, but didn't change the brain; they only talked to it. We could imagine, in the not too distant future, having a better understanding of how the brain works internally (this is an area of science that's moving at least as fast as nanotechnology, by the way). Then we could inject signals, not simply into the incoming and outgoing nerves, but inside the brain itself. Instead of seeing the answer to be read in your visual field, you'd "remember" it. Your imaginary assistants would "read your mind" and not need to be told what to do.[1] You could look at a pen and know that it was exactly 152.4 millimeters long the way you look at it now and think it's about 5 or 6 inches.

Moving in a different direction for a moment, we could take the brain and replace it with a machine that duplicated the functions of the neurons, together with other circuitry that simulated the effects of endocrine balances and other chemical signaling pathways. The machine could be smaller, lighter, and much more durable than the actual neurons of the brain. More important, it could be a lot faster. In the same volume as the human brain, you could put somewhere between a thousand and a million times the raw computational power. Now your imaginary friends are not so much a family as a city—all still intimately aware of everything you want, but in a complex organization with highly specialized members who can work at a level of detail that you'd never have time for. And it's all inside your head.

You might now be fit, for example, to live in a tribal everybody-knows-everybody community of two hundred thousand people, instead of just two hundred. You might be able to assess the risks, gauge the advantages, and intelligently design and implement complex technologies and social systems. With a little extra effort, you could probably set the clock on your VCR.

WHY BOTHER?

I frankly can't understand the technophobe's point of view. I've tried, really. One book that was reccomended was by Bill McKibben, author of *The End of Nature*.[2] His latest call to give up technology, in this case the use of biotech to augment humans, is called *Enough: Staying Human in an Engineered Age*. McKibben takes this view of improving intelligence:

> Of course, the problem with arms races is that you never really get anywhere. If everyone's adding 30 IQ points, then having an IQ of 150 won't get you any closer to Stanford than you were at the outset.[3]

Getting into Stanford is far from the best reason to want to change your IQ, or that of your children. I would be much happier in a world where everyone else's IQ were thirty points higher, even if mine remained the same. Consider that crime, which is strongly correlated with low IQ, would virtually vanish.[4] For the average person, thirty IQ points would move them from the category where they need training for a new job to the one where they can pick up a book and do it themselves, where they are much more likely to understand the consequences of various life choices, and thereby would be better off in the long run. Public discourse might be raised from its current level of least-common-denominator banal, stupid, emotion-laden sound bites to slightly more informed discussion. Not, of course, because the politicians had gotten any closer to being moral animals, but simply because the least common denominator had become more intelligent. I might find Bach instead of Beastie Boys playing on diner jukeboxes. I might be able to talk to my barber about thermodynamics instead of the temperature.

McKibben doesn't really find too many specifics to argue with in the prospect of augmenting humans. This is surprising, given the general tone of the book.[5] In many cases he simply displays the ideas of the transhumanists (proponents of technological self-improvement) often quoting people I know and/or whose opinions I respect, leaving them hanging there as if they were horrific enough all by themselves.[6] Perhaps they are, to him. A couple of problems that he identifies do deserve some thought, though.

First, as a marathon runner, he bemoans possible physical enhancements that would make marathon running easy. Where would be the joy of accomplishment, the feeling of having measured yourself against a challenge, if there were no challenge?[7]

I find this specious. Marathon running is a game. It's not a true challenge, but one that is made up for people to measure themselves against. A true challenge, for a true human, is one where success or failure makes a difference. Enhanced humans can play games, too, and they will be just as meaningful as today's games are. You'll be measuring yourself against someone else who's enhanced. Car racing is a widely appreciated sport. So are women's tennis, soccer, and basketball, as is Little League baseball and the Special Olympics. The fact that there are different classes of competition doesn't make the game less rewarding.

On the other hand, true satisfaction when you look back over your life doesn't come from having played games. It comes from having done things that mattered. Yes, we can meet with triumph and disaster, and treat those two imposters just the same; but when life is on the line, triumph is better. McKibben seems not to realize that there are greater challenges, worthy of nanotech (or biotech) capabilities.

Oddly enough, one of these challenges squarely answers the concerns of McKibben's earlier book, *The End of Nature*. There he worried that the advance of technology was making the Earth into a managed park, even if we do decide to take care of it. No more truly wild lands, nothing beyond the reach of grasping bureaucrats or greedy developers.

This happens to be a concern I share. But I also have an answer. The universe at large is wild, natural, and forever (yes, forever) beyond the ability of humans to fence in, domesticate, and tame.[8] Of course, from where we sit on Earth today, it doesn't seem like a challenge; it seems impossible. For it to be a decent challenge, we've got to get out there and measure ourselves against it. As nonaugmented, biological humans, we're fish looking up at dry land. But our forefathers, the actual fish, managed to conquer the actual land. We can do as well.

The other concern in *Enough* worth considering is the notion that parents might program their children off in different directions and humanity will lose a common basis. For example, a musician might enhance the musical skills of her child but also strengthen the connections from music to feelings so much as to alter the child's basic outlook. Meanwhile a cleric might program his son for religious faith. Ultimately we'd get people who simply couldn't understand one another. Sort of like the English and the IRA, the Israelis and the Palestinians, or Bill McKibben and yours truly.

Consider: You have the new Humanity 1.5 Control Panel installed. You can pop up the panel, a virtual reality gadget, into your field of view. It has a bunch of gauges, buttons, sliders, and so forth familiar from software applications. But these say things like Height, Weight, Hair Color, a selector menu for Racial Type, a slider that varies between Fine Motor Control and Gross Motor Control as your optimum coordination regime. Scroll down some and you find Intelligence, Personality Type, Interpersonal Sensitivity, Irritability, Ambition, Gregariousness, and so forth.

Twiddle the dials some. You get taller or shorter, fatter of thinner. Except for these gross physical attributes, though, you won't notice much difference until you interact with people or tackle some problem. I don't see any major objection to letting people have this kind of synthetic self-control, in general, provided some common sense is used.

OK, fine. Now you get panels for your kids. Who gets to set them? The kids? You? Schoolteachers, guidance counselors, bureaucrats? Who gets to play with *yours*? Policemen? Judges? Anybody who wins a lawsuit against you?[9]

THE PROGRAMMABLE ANIMAL

> I'd encountered the idea that we were all cyborgs once or twice before, but usually in writings on gender or in postmodernist studies of text. What struck me in July 1997 was that this kind of story was the literal and scientific truth. The human mind, if it is to be the physical organ of human reason, simply cannot be seen as bound and restricted by the biological skinbag. In fact, it has never been thus restricted and bound, at least not since the first meaningful words were uttered on some ancestral plain. But this ancient seepage has been gathering momentum with the advent of texts, PCs, coe-volving software agents, and user-adaptive home and office devices. The mind is less and less in the head.
>
> —Andy Clark, *Natural-Born Cyborgs*

The problem is that McKibben is a couple of million years too late. Starting (very roughly) two million years ago, we begin finding in the fossil record an artifact that archeoanthropologists call the Acheulian handaxe. In fact, it's neither particularly Acheulian nor a handaxe, but the name has stuck anyway. What it actually is, is the granddaddy of all technology.

The Acheulian handaxe was made and used by our ancestors for something like 1.8 million years. No other piece of technology is even close. When our ancestors began using it, they were *Homo erectus*; by the time people had quit making it, they were *Homo sapiens*. We shaped the stone, but the stone also shaped us.

The handaxe puzzled anthropologists for many years. It is a piece of carefully chipped rock, flat, with a sharp edge, that has an outline like a pointed egg. The trouble is that it's sharpened all the way around; if you used it for a handaxe, you'd cut your hand. And there's no need to go to all the work to make a handaxe symmetrical.

Thanks to some inspired detective work by William Calvin, we finally have a good idea how the handaxe was used. It was a killer Frisbee, to be thrown at animal herds when they were tightly packed at water holes. It works by a complex interaction of phenomena, including aerodynamics of the particular shape, the nature of animals' reflexes, and herd reactions to hunters. In particular, it works, often enough, when you aren't nearly a good enough rock thrower to hit a specific animal in a vulnerable spot, but can hit the herd somewhere. Essentially the handaxe, however thrown, will flip vertical in the air, the sharp edge will hit the animal, and the point will catch instead of rolling off. The animal will flinch and trip as the herd stampedes, leaving it knocked down, trampled, and dazed enough that the *Homo erectuses* could run up and grab it.[10]

This took some discovering. However, it didn't get rediscovered again by each new hunter over the course of nearly two million years; it got copied. Indeed, the handaxe itself almost certainly evolved over many generations from hunters seeing that some stones worked better than others, and trying to chip bad stones till they looked like the good ones. So while *Homo erectus* was evolving hand-eye coordination and becoming a better stone chipper and thrower, he was also evolving the ability to imitate—and becoming a substrate.

In other words, the human brain became not a machine for controlling a body, as in an animal, but a machine for running programs, some of which controlled the body, but which could also speak and understand and invent and experiment and teach and learn.

By the time we became *Homo sapiens*, we had developed our coordination, which included our brains, to the point that an experienced hunter could throw a spear thirty yards and hit an animal in a particular spot with enough force to deliver a debilitating wound. We had also developed rich, structured, symbolic language, using the same part of the brain as for sequencing complex hand motions. We are now evolved, designed if you will, to copy ideas, techniques, and knowledge from each other. We are meme machines.[11]

It's worth clarifying at this point that a meme is a pattern of information considered as a replicator. Memes have been wildly misunderstood and misrepresented by a number of authors, called things like "virus of the mind," and so forth. Memes, including the knowledge of how to make and use an Acheulian handaxe, do spread by being copied. But it's not as if that were different from any other knowledge the mind holds. Everything in there got copied from parents and other members of the community. Being an informational virus is the merest fringe of what memes are. They are the software of the mind. All of it.

It's true that a good part of what it means to be human is given us in our genes. That is also, to a 99% overlap, the same part of what it means to be a chimpanzee. That tiny difference is the supercharged idea-copying machinery. (Chimps copy techniques, but with about the same facility that humans mentally multiply nine-digit numbers. It can take a chimp seven years to learn to crack nuts between rocks by watching other chimps do it. We're *good* at copying.) All our basic emotions, patterns of behavior toward others, even politics, are basically chimplike.

That one difference is huge, though. People are an idea substrate. We are built by taking a young, blank organism, and filling it with ideas, techniques, attitudes, concepts, knowledge. We are just as much formed by our cultural heritage as by our genetic one. Indeed it is not going too far to say that in a very real sense, we *are* our ideas. If you have an internal conversation with yourself, conscious creature that you are, you are using concepts and abilities that were learned as part of your culture. A feral human cannot use language and may not even be conscious in the way we are.

What price modifying humanity if "what we are" is really "the software"? On one point, I find myself in agreement with McKibben again. A man who's been genetically programmed to be pious, unable to doubt his faith, has been reduced to a subhuman state. His hardware has been crippled so that it cannot run the full range of software. But so has the child raised by radical environmentalists and trained to hate the idea of technological progress. Both are reversible, by the same genre of techniques with which they were done; but in neither case would the individual want to be deprogrammed.

What's more, our genetic component isn't a specification for a brain, anyway. It's a specification of an algorithm for building a brain that is heavily dependent on the environment the brain develops in. It's been shown recently that watching television as a baby can greatly increase one's chance of an attention deficit disorder. It's not hard to see why: watch a TV for a few minutes and count the number of scene shifts. Especially in commercials, it's rare for the camera to point at the same thing for much more than a second. When a brain forms under such a torrent of continual change, it will tend to be better at quickly parsing newly flashed information, but worse at deeply analyzing something that remains the same for long periods. TV is packed, intentionally, with as many attention-getting phenomena as possible. To a mind formed on TV, the attention-getting signals in the real world are sparse and weak.

In other words, we're already rewiring our kid's brains simply by the constructed environments we live in, as well as programming them with everything we teach and everything they watch us do.

In my own concept of what it means to be human, copying the software to a new substrate can preserve the essence of being human, if the new substrate is able to run the software the way the old one would. This point is very important, more so than realized by some transhumanists, I think. At the current state of knowledge in cognitive psychology and neurophysiology, there's lots of stuff we don't have a clue about. We do know that, for example, emotions and the sensory feedback from the body are important; the human mind is first and foremost a body controller, and second and almost as important, an interacting unit in social groups.

So it will not be an easy or a simple job to build a compatible substrate. It will involve more than simply simulating the neurons; some kind of (at least virtual) embodiment appears necessary. However, it is by no means impossible. We have a lot to learn, but we're learning fast in the areas we need.

Now we can give some considerably more enlightened guidelines for staying human in an engineered brain. The prime notion is from computer science: "upward compatibility." In your new hardware, you must be able to think all the same thoughts, have all the same feelings, experience all the same stimuli as in the old. Only then can you talk about expanding the repertoire. One obvious way to go is to be able to think more of the thoughts, know more of the knowledge, and have more of the experiences that humans have always had. More memory, more integrating circuitry of the same type, faster processing. Simply being able to see things from more points of view, to use more of the knowledge available to make decisions, could make us better people— wiser as well as more capable.

Another really major improvement that would go a long way toward making people a lot happier is to add cognitive abilities that allow us to make sense of the modern, formalistic, mechanical, engineered world with the same instinctive ease that we evolved to handle the wild savanna. You can pick out a tiger slinking behind a screen of brush with a single effortless glance. The data processing involved in putting together the hints of light and shadow, patterns of motion in the visual field, and comparing the result with an enormous stored database of possibilities, is still beyond our best computer systems. It is considerably more computation than would be involved, for example, in filing the most complex income tax return. But we live in a world of taxes, not of tigers. There's no reason, in principle, why the relatively simple information-processing tasks we face today can't be as effortless as the considerably more complex ones we already do.

Obviously more efficient, you may say, but happier? Leave your office, the boxes of steel and concrete, and stand in a park, a field, a forest. Watch the clouds, deer wandering by, or a sunset. Look up and see the scattered diamond dust of the stars in the night sky. These things are beautiful and restful to us because we're evolved to appreciate them. The good things in our created world should evoke the same sense of rightness and satisfaction.

But probably the most important class of improvements will be an ability we might call telemagination. Consider having the reverse of vision: the ability to create pictures instantly of whatever it is we're thinking of. Our computers already do this. Then transmit the picture, along with imagined sounds and other sensory impressions, to someone. In real time, it's a movie, produced as effortlessly as talking is today. The really revolutionary aspect of such a skill would be to open up the bandwidth of person-to-person communication. It might be possible—this remains to be seen—to compact some communication into a more condensed semantic form; but failing that, movies are pretty good.

To see the difference between people today and people with telemagination, think of the difference between a PC and a PC connected to the Internet. Or the difference between *Homo erectus*, who did not have language, and *Homo sapiens*.

Adding a few new cognitive abilities, and a lot more horsepower, to the old noggin will, yes, make us into something not human as we conceive it today. So did using the handaxe change our ancestors. We still have the *Homo erectus* mind, and indeed the chimpanzee one, as a vital part of our basis; but we have added to it, in what I believe is a good way. So will our descendants, and possibly ourselves, add to what we are today. Surely it is the most arrogant conceit to think that we, now, are the apotheosis of evolution, the best that can be, intellectually or morally. Only a fool believes himself unimprovable.

We have improved ourselves for millions of years by adding ideas to the bundle of information we consist of. Our genetic heritage of information, encoded in DNA, is given almost entirely to each of us, indeed ten trillion copies. Each cell has the whole thing and can use whatever parts of it are appropriate, to the limits of its abilities. It's a good model.

We've gotten to the point where our culture, the informational heritage of humankind, is a staggering overload for any person to learn. It's not that our minds couldn't handle it—after all, it was all thought of by people in the first place. Our ape brains just don't have the horsepower. Our informational selves have outgrown the substrate, and we are in serious danger of falling behind our own machines. It's time for an upgrade.

We are like the bound feet of Chinese women in historical times, living in a tourniquet, unable to grow. Our minds could grow naturally to follow our culture, art, and science if only we didn't have to cram it all into a few pounds of gray matter evolved to steer a hunter/gatherer around the plains of Africa. If we can only learn to stop identifying ourselves—the patterns of our thoughts and feelings—with the lump of matter those patterns are formed in, we can move into more spacious accomodations and have room to grow. Then we can emerge from our cocoon and stand on our own feet among the thinking machines of the future.

CHAPTER 19

THE HUMAN PROSPECT
Living Happily Ever After

Da steh ich nun, ich armer Tor!
Und bin so klug als wie zuvor.

(And here, poor fool! with all my lore
I stand, no wiser than before.)
—Johann Wolfgang von Goethe, *Faust*

One way to sum up nanotechnology is that it will make matter into software. Nanotechnology is, of course, hardware, but it has many more features in common with today's software than today's hardware. Fifty years ago, software was an arcane art practiced by an elite priesthood in rare, expensive, and impressive settings. Today, everyone uses it, be it on a computer, cell phone, or ATM. Given the Moore's Law expansion of processing power, what software can do is limited only by the imagination.

Software is incredibly intricate and complex. There is an enormous amount of technology that goes into software in and of itself: everything from the low-level details of a given computer's instruction set, to the operating system and its interfaces, to compilers and software development environments, to computer science with its mathematical analysis of algorithms. However, for any given application, there is an entire body of knowledge about the end user's field of interest that has to be added. For a word processor, it's knowledge of letters and words and fonts and formatting. For an engineering CAD system, it's knowledge of three-dimensional geometry, measurement systems, drafting techniques, and the like. And on and on, for Web browsers, airline reservation systems, geneology database programs, the works.

In just the same way, nanotechnology will have a core of techniques required for any application. But for each end purpose, more and specialized knowledge will be required: biology for nanomedicine, aeronautics for flying cars, civil engineering for structures, and so forth.

If you use a computer nowadays, what you actually interact with is the software. All the things you see: windows, pictures, text, menus, cursors, icons, and the like, are created by the software design, not the physical computer. The same physical computer, with different software, could just as easily give you a completely different interface. Many applications do this: a flight simulator has you interacting with a completely different set of concepts than a recipe manager.

In the physical world, of course, the interfaces to an airplane and a kitchen are different, too. The cockpit controls look and act nothing like your range, sink, and refrigerator. But the feature of software that will carry over into nanotechnology is that the world you experience with software is completely made up. The only limit is the designer's imagination. With nanotechnology, the world you experience—not just patterns of light on the screen but the whole physical world—could be entirely made up, as well. As with any new, glitzy technology, this seems likely to be overdone at first, but soon, let us hope, wiser counsels will prevail, and reasonable, moderate uses of technologies such as Utility Fog will be the norm.

A WORLD OF YOUR OWN

For two million years or so, since those first chipped stones, advances in technology have redounded to the advantage of those who had it. Faced with an overwhelming and often hostile natural world, our ancestors needed all the help they could get. Technology extended their power and gave them more mastery over that world.

Since the invention of agriculture, real property, and cities, however, there has been a subtle reversal in the trend, which has come to a head in the past few centuries. Once nature is essentially mastered, extending one's grasp can only encroach on the lives of other people.

On the average, nobody can win. Life comes to resemble a zero-sum game in many ways. With an increasing population, it's worse, since each person's share shrinks.

A zero-sum game is a recipe for unhappiness and strife. We are built, psychologically, to enjoy doing good things for ourselves and for others. In a zero-sum game, you can't do good for anyone without doing harm to someone else. The very aspect of the Prisoner's Dilemma that allows cooperation, and thus morality, to evolve, is that it is not a zero-sum game. Without the evolutionary pressure of the Prisoner's Dilemma dynamic, the basis of altruism and morality disappears and they begin to erode.

Humans always will—and should—compete, or they wouldn't be human. But in a static world, empathy and kindness are repressed by the lack of opportunity. Add to this the sense of frustration created by a serious discrepancy between what one could do, in the absence of others, and what one is allowed to do. In reality, in order to gain in such a world, people will begin looking for ways to demonize others, so as to be able to hurt them with a clear conscience. Blame and injury breed deep-seated hatred and strife.

Such a world is an evil place, but it is one we are heading toward all too clearly, if the conventional vision of limited resources "just one Earth" holds sway. Technology both helps and hurts. When you sit in a traffic jam, annoyed at the other drivers, it seems the automobile is the root of the problem. But it's really people, as the pedestrian traffic jams in China's cities demonstrate.[1] The auto is well matched to population densities outside of cities. But the auto, or any other technology that extends one's range, exacerbates the problem when that range overlaps other people's. People evolved to live in an uncrowded world and to associate on a more voluntary basis with fewer other people than they typically do now.

Technology helps by increasing total wealth. This makes the world less zero-sum in some ways. Increasing the total stock of value means you don't have to rob Peter to pay Paul.

One reason computers have become so successful over the past decades is that in a very real sense, they expand the world. With a computer, there are more things to do, more items of interest, and more people with whom you can associate or not as you please. We may be forced to rub elbows with strangers in the physical world, but there's plenty of room in the virtual one.

Nanotechnology can expand the world in many ways beyond the assisted sublimation of the computer. There really is plenty of room at the bottom. Acting through a virtual reality link, a two-millimeter humanoid telerobot would find a ten-acre field as vast and daunting as Henry Stanley and David Livingstone's Africa, and filled with stranger, more varied, and more dangerous creatures. Or you could live in literally limitless virtual worlds indistinguishable from physical reality to your senses. You could do more real, physical things in less space, including those nanomachine shops in your design workstation. And you could physically move away, living in remote areas on Earth or in outer space.

ELBOW ROOM

> The wild deer, wand'ring here & there,
> Keeps the Human Soul from Care.
> —William Blake, "Auguries of Innocence"

A well-used basketball is covered with a layer of sweat, oils, dead skin cells, and living bacteria. This layer is thicker on the scale of the basketball than the layer of life, the biosphere, is on the scale of the Earth. This layer contains not only all intelligent life, but all known life of any kind, in the universe.

The biosphere seems huge and limitless to an individual human. For humankind as a whole, it's beginning to get a bit tight around the elbows. Edward O. Wilson, the eminent biologist, suggests that *Homo sapiens* may well already have exceeded the long-term carrying capacity of the biosphere in the late twentieth century.[2] It's completely clear that we're well over the capacity of the environment to feed us, absent large-scale agriculture. And that means agriculture with irrigation and synthetic fertilizers—simple farming won't do it.

To get an idea of how precarious a position we've gotten ourselves into, simply consider what would happen if some new disease did to wheat and rice what the Dutch elm fungus did to elms in the United States since 1930. It wouldn't really matter whether the disease were human-made or entirely natural.

The human species is vulnerable because everything we are, need, and value is here in this vanishingly thin film of life on one planet.

And yet we are increasing the load we place on the biosphere, both in resources per person and in the number of people.

How many people can the Earth hold? With nanotechnology, quite a few more than without it, just as agriculture itself increased the practicable population density. According to the US Census Bureau, the current population density of the United States is about eighty people per square mile. That's already too many for each person to have a ten-acre estate (and the United States is among the more sparsely populated countries).[3] The most densely populated area in the United States is Jersey City, New Jersey, with 13,044 people per square mile. (New York City has only 8,159, and the average nation-wide is under 1,000.)[4] If we were to populate the entire land surface of the Earth to Jersey City densities, it would hold about 750 billion people, or 125 times its current population. There would be no room for fields, farms, parks, forests, or anything; all food would have to be produced in the oceans (or synthesizers). Note that that population is only seven doubling times away from our current world population of 6 billion.

With nanotechnology, the prospect is less hellish. We can build mile-high towers covering only 1% of the land surface and house five times as many people while leaving 99% of the surface in a natural state. This would give us two and a fraction more doubling times. Rather than agriculture, nanotech recycling would stand in for the carrying capacity of the biosphere. At this point, though, living in one of these giant hives, you'd depend on virtually the same technology that would be required to live in space, and visits to the real biosphere would be strictly regulated to prevent the swarming people from trampling it flat. And we're still in trouble at ten doubling times.

How long is a doubling time? Recent US growth rates have been about 1% per year, which gives a doubling time of seventy years. Historical growth rates have been over 3%, for a doubling time of twenty-five years. It is generally believed that a rising level of afflu-ence is correlated with declining growth rates, and that's all for the good. However, nanotechnology will most likely throw a monkey wrench in that machinery. Nanomedicine will very likely extend life-times significantly, even indefinitely, and preserve youth, vigor, and childbearing ability for a much longer term. This means that people will have the ability to have a career, retire with a comfortable pen-sion—and then have a family.[5] (People do seem to enjoy their grandchildren more than they do their children.) What's more, the mortality column in the population tables will decline toward zero. Census Bureau projections for 2050 give, somewhat optimistically, a birth rate of 1.4%, a death rate of 0.98%, and a net immigration rate of 0.25%, for a growth rate of 0.67%, and a doubling time of over a century.[6] But if the death rate goes to near zero, and childbearing age becomes twenty to one hundred instead of twenty to forty, we might easily get a 3% growth rate again. Worldwide Jersey City by 2200!

How about restricting birthrates, as China does? The picture is relatively stable with one kid per couple, but now there's a whole new reason to want to go to space: to have babies. The main problem with staying on Earth is that it's squarely against both our deepest instinctive drives, to stay alive and to have children. We've either got to stop making babies or start making land. The latter seems more preferable.

It's worth pointing out that land, in the form of O'Neill colonies, can be built in such a way as to accomodate extensions of Earth's biosphere. Population pressures on Earth put our values for babies and wildlife directly at odds with each other. The asteroids alone could provide the material for colonies with twenty thousand times the land area of Earth. No species need be crowded out of existence. Country or wilderness living can continue to be an option, although a full-fledged rotating, shielded colony will always be more expensive than a simple house in space.

If there's only one Earth, and everyone lives here, and everything that humans do is done here, retaining significant chunks of the biosphere as nature preserves will become harder and harder as time goes on. If we move into space, however, the Earth will ultimately be less than 1% of 1% of the total livable area humans have available. It seems much more likely that a significant portion could then be set aside and a large proportion of living species saved from extinction.

In the long run, of course, we'd better be living in other places than the Earth, or the next dinosaur-killer asteroid will drop the curtains on our little show. The chance of this happening in any given year, however, are about the same as your chances of both winning the lottery and being killed in an auto accident in the same year. People being what they are, I wouldn't count on any major projects being undertaken on the basis of that risk any time soon. Still, we'll be safer as a species in the long run with our eggs in several baskets.

There are other dangers to worry about, though. It's always been the case that a major plague could wipe out a large fraction of humanity. We've grown complacent with modern antibiotics, and we may even have a right to be, as nanomedicine advances. However, the increasing sophistication of biotechnology makes it likely that quite nasty plagues could be created from scratch. This could happen either as a purely malicious act by a madman or terrorist or by semiaccident. Consider a scenario where delinquent youths release killer plagues for a thrill, believing that no major harm will be done because of advanced nanomedicine. This would correspond to today's "script kiddies" who write computer viruses. Then one of the plagues evolves to be just a bit trickier than the current defenses can handle.

Replicating nanomachine infestations are also a possibility, though the biotech ones will be possible sooner. And good old-fashioned nuclear terrorism won't go away—nor will state-sponsored nuclear war. Frankly, I don't expect to see any of these threats taking a sizable bite out of the human race in the next century, but I do expect to see increasing restrictions of freedom of activity, movement, inquiry, and information in an attempt to forestall them.

A BALANCE OF POWERS

> Certainly it seems that nothing could have been more obvious to the people of the earlier twentieth century than the rapidity with which war was becoming impossible. And as certainly they did not see it. They did not see it until the atomic bombs burst in their fumbling hands. Yet the broad facts must have glared on any intelligent mind. All through the nineteenth and twentieth centuries the amount of energy that men were able to command was continually increasing. Applied to warfare that meant that the power to inflict a blow, the power to destroy, was continually increasing. There was no increase whatever in the power to escape.
>
> —H. G. Wells, *The World Set Free*

There is a natural balance in the capabilities for destruction and transportation in any given level of technology. A man with a sword might slay a family unawares in his village but he's unlikely to best five other men with helmets and shields. A man on horseback can chase down peasants on foot, but others on horses of their own escape him. Chemical explosions confined in steel cylinders give the gun a great destructive range, but in the automobile they give people the ability to live out of gunshot from each other and still engage in daily commerce.

The most destructive technology of the twentieth century, the nuclear explosive, could similarly have extended the range of practical transportation. The NERVA[7] and ORION[8] rockets could have opened up the solar system, and nuclear power plants could have powered colonies on places like the Moon and Mars, and operations like O'Neill's lunar mass driver. This was not done, of course, and the impact of nuclear technology has remained strongly biased toward the destructive. This seems to be typical of technology that remains largely in government hands, and the phenomenon has contributed to a perception of technology as more dangerous than good in the twentieth century. There's a pernicious feedback effect: the perception of dangerousness feeds the restriction, which in turn prevents the general public from gaining experience and finding beneficial uses for the technology. Thus most uses found are for military applications, feeding the perception of dangerousness.

It's not all a question of technology, of course. The bounds of the Earth put us in a boxing ring that is large compared to horse or automotive technology, but small compared to rocket and nuclear. With nanotechnology, it will shrink again. Biotech weapons are the first where a single person can carry something that would threaten a substantial fraction of humanity. They will not be the last. Nanotechnology gives us the ability to move outside the box. It makes possible single-family dwellings in space that are comfortable, self-sufficient, and hundreds of miles apart. A terrorist with a hydrogen bomb, facing such a spread-out population, would be reduced to the destructiveness of the man with a sword.

There are several implications. First is the obvious: given someone with such a weapon and the will to use it, he imposes a risk that would be cut by a factor of a million. That means that such an act would not have anywhere near the political effect of Earth-based terrorism, so the motive to do it in the first place would evaporate. Life in space could be both free and safe. You could trust you neighbors even if you didn't watch them constantly. A cluster of homes in a quintillion cubic miles, tiny on the scale of the solar system, could easily be a more tightly knit community than a twenty-mile square of city on the Earth's surface.

Back on Earth, we're all living in the same soup bowl. Any accidental or purposeful release, whether replicator or merely poison, pollutes the one atmosphere we all breathe. Nanotech defenses and countermeasures are certainly feasible and, I believe, will prove to be effective; but once you are wearing a skinsuit and breathing filtered air anyway, the jump to a spacesuit doesn't seem all that great.

Finally is simply the question of the quality of life. People will usually take the choice of more room over less, all other things being equal. They will choose safety over danger, freedom over restrictions. They will have families. They will seek affluence over poverty.

A large part of affluence is energy. Current global energy consumption is about 2 kilowatts per person. This already has some effect on global climate. The effect can be reduced by moving away from fossil fuels, but we can increase energy use by only a factor of ten, at most one hundred, before the energy itself—even if its generation is absolutely clean—begins to affect the climate susbtantially.[9] A large part of this headroom will be consumed merely in raising the existing population to Western levels of affluence. (Energy consumption in the United States is about 12 kilowatts per person.) Remember that it's rising affluence that demographers are counting on to ameliorate population growth rates.

If we were in space using solar power, and the population were one thousand times what it is now, and each person were using one thousand times as much power as we do now, we'd still be using less than a millionth of what the Sun puts out. The streets of space are paved with gold—golden sunlight.

Perhaps the most important part of the quality of life is that people need a challenge and an environment in which their exertions can affect their prospects, and those of their children, for the better. Humans need to face human-sized problems. Too hard and the result is continual failure and a broken spirit. Too easy, or no challenge at all, and the result is boredom, apathy, and ultimately, random acts of irrational destructiveness. With nanotechnology, Earth is just too small a sandbox. We must now step out and measure ourselves against the stars.

MORALITY AND CENTRALIZATION

> In my day a reporter who took an assignment was wholly on his own until he got back to the office, and even then he was molested little until his copy was turned in at the desk; today he tends to become only a homunculus at the end of a telephone wire, and the reduction of his observations to prose is farmed out to literary castrati who never leave the office, and hence never feel the wind of the world in their faces or see anything with their own eyes.
>
> —H. L. Mencken, *Newspaper Days*

The astute reader will have noticed that in this book, I have tended to draw a picture of nanotech applications, where possible, as coming in a world where individual control and responsibility hold sway, and the benefits are aimed at individuals. The twentieth century saw a widespread trend toward another extreme, where responsibility moves from individuals to formal organizations. It might be reasonably asked whether the trend will actually stop, or whether we will ultimately wind up essentially pets of the government, with the right to choose which flavor of ice cream we want but not to do anything that makes a significant difference in the world.

One thing that is happening is that there are more people, but less work. As manufacturing and other systems become more capable and efficient, the jobs are spread more thinly, making each person responsible for less and less. Such specialization makes people more efficient but has a tendency to shift responsibility, power, and rewards, both psychic and material, from the individual to the organization. Ironically it was Karl Marx who was first noted for discussing this phenomenon, which he called alienation.

Two other obvious pressures are behind the trend. First is simply that organizations will take power when they can. This is a purely evolutionary dynamic, since organizations that take power will tend to displace ones that don't. Second is that people will use organizations to shield themselves from responsibility wherever possible. In the long run, of course, when you avoid the responsibility for something, you give up the corresponding power as well. Thus things that used to be in the purview of personal common sense are now centrally controlled and regulated.

It is not as if any common sense were directing the central regulation. The US government regulates the number of ounces of water you can use to flush your toilet, but sells irrigation water for less than a penny a ton to rice growers in semidesert areas. The FDA causes the deaths of thousands of Americans every year by refusing to certify drugs that are widely in use in Canada and Europe, in the name of safety.[10] Meanwhile, federal agents arrest people in California for using marijuana for medical purposes (to relieve glaucoma, for example), although such uses are explicitly legal there. Everyone has his favorite example of the foolishness of centralized control. One of mine is from World War II: The Allies wanted to impede the German war machines by constricting the supply of ball bearings. So they sent a bomber raid to demolish the headquarters of the nationalized German ball bearing industry. Ball bearing production tripled.[11]

In some areas, such as economics, it's come to be understood that central control is a disaster. Economies work only because of the widespread information gathering and decision making done by everyone all the time as buyers and sellers. In 1900 serious, thoughtful people could believe, and many did, that rational, scientific control of the economy could be more efficient and less wasteful than the helter-skelter hodgepodge of the marketplace. But after some large-scale experiments, such as the Soviet Union, a deeper understanding was gained.

Nanotechnology will bring the issues of personal responsibility to a head. For example, should people be allowed to make drugs and weapons in their home synthesizers? How about two-gallon-per-flush toilets? Medical equipment for use in emergencies? The crucial issue is that nanotech and AI could make it possible for an extension of the state to be watching every person every minute and controlling everything they do.

Humans were evolved to be semiautonomous. We prefer to live in small groups but are, in our natural state, quite capable of existing as individuals or family units. Large groups of hunter/gatherers depleted game and forage stocks and were not favored. With agriculture, more substantial groupings evolved. The largest concentrations of people, even in antiquity, were associated with considerable technology: the aqueduct and sewer systems of Roman cities, for example. With the coming of the industrial revolution, however, the "society as machine, people as parts" model really took hold.

It might reasonably be projected that nanotechnology, as an

extension of the industrial revolution, might simply continue the trend. But that's not necessarily the case. Today's economy is highly interdependent: a complex product may have parts or materials from a hundred countries. Nanotechnology can reverse this completely. Your countertop synthesizer will have more moving parts than there are in all the machines on Earth today. The trend to complexity can continue, but at the same time we have the opportunity, if we desire it, for a level of autonomy, independence, and freedom, equivalent to that of the hunter/gatherer.

THE NEW FRONTIERS

Nanotechnology offers frontiers in many directions that were not available before. The solution to crowding and resource depletion is clearly to move into outer space. If you look at it from a broad perspective, space is the real universe and the Earth is a microscopically small bubble in a vast sea of room and resources. It's ridiculous to stay here when we've so clearly outgrown our original niche as a bipedal ape.

Space sounds far-fetched because the practical difficulties are substantial. A shuttle flight requires thousands of technician-years of maintenance for preparation. How can something like that ever be economical? Perhaps a final comparison with computers will help. When you open a document in a word processor, there's generally at least a one-second pause before the document appears on your screen. In that second, a modern PC will have done a billion operations as complex as multiplying ten-digit numbers. Just fifty years ago, a clerk doing just the same bookkeeping or arithmetic would have taken at least a minute for each one. Thus the one-second pause of your word processor represents nearly two thousand solid years of clerical work.

As nanotechnology matures, the same ability to throw what would have been enormous efforts at the most trivial problems will come to the physical world as we have in the software world now. Living in space is dangerous and prohibitively expensive with current technology; it will be cheap, easy, and safe with advanced nanotechnology. We really can simply leave our problems behind.

UPLOADING

> Why, if the Soul can fling the Dust aside,
> And naked on the Air of Heaven ride,
> Were't not a Shame—were't not a Shame for him
> In this clay carcass crippled to abide?
>
> —Omar Khayyam, *The Rubaiyat*

Uploading means copying your brain and as much of the rest of your nervous system and bodily reactions as necessary, in software form, into a computer or other substrate that gives you more speed, processing power, memory, longevity, and room to grow than the original biological equipment. Some people might want to exist as individual robots, while others would prefer to live in the same huge computer, with much higher communication rates between them. This is obviously a choice not everyone has to make the same way. We can expect to find far-flung scatterings of rugged individualists and centralized concentrations of the gregarious. For people living in the more crowded of accomodations, it would make sense to live primarily through virtual reality. Interactions with the physical world would be by virtual reality links to android robots, or for extended excursions, mounting the brain in one.

The physical human brain is slow and fragile, and has idiosyncratic dietary requirements. Better, perhaps, to reimplement it in a software substrate. This would have some significant benefits. You could download yourself into a robot instead of having to use a network connection or go to the bother and danger of having it physically carry your brain around. You could further avoid danger by making backup copies of yourself, in more than one place. You could avoid tedious journeys by transmitting yourself as data from one robot to another. When not using a physical body, you could run, connected to virtual reality worlds, on powerful stationary processors thousands of times faster than physical-world subjective time rates.

On the other hand, it will probably take a while to convince you that once you get your brain copied off into software, it's really still *you*. After all, we've pointed out that atom-for-atom teleportation by analysis and reconstruction needs data rates too high for even likely nanotech networks. The difference lies in the fact that the mechanism of the brain that gives rise to thought, consciousness, memory, and so forth is almost certainly at a higher level than the molecular. Most of what goes on in brain cells is the same stuff that goes on in muscle, skin, and fat cells: DNA and hormones regulating the activities of ribosomes and metabolic pathways, and mitochondria providing ATP to power those and other activities.

It's very likely that the level of activity that supports our consciousness is the firing, synaptic transfer, and internal processing of the electrical signals in the neurons. It's certainly true that they are crucial to all the brain, sensory, and motor functions that have been studied. Just as certainly, other inputs and processes are involved, but they are not mysterious, merely as yet incompletely understood. It's only reasonable to assume that the same higher-level processes could be supported by a different set of fuel-burning, power-generating, and construction machines.

This is where most people's intuitions about the process break down. "Okay," they say, "I'll grudgingly grant that you could build a machine that would act like a brain, and maybe even act like *my* brain, but it wouldn't really be *me*. If you took my brain apart to create a software version, you'd have killed me and created a new robot that imitated me in a ghoulish fashion."

One way to get around the intuition is the scenario by Hans Moravec in his book *Mind Children*. He imagines that you can be conscious while the conversion is being done, and that it is incremental, a small clump of brain tissue at a time. The process is also reversible, so that you can experience being in the computer form, return to your own brain, and remember the entire episode. If, when this happens, you experience continuity of consciousness and your memories are consistent, it becomes harder to argue that it wasn't really you. The crucial point is that, having experienced it, you would *believe* it was really you.[12]

Not everyone's intuition is the same, of course. Many people believe in reincarnation, and a widespread audience was capable of the suspension of disbelief to make *My Mother the Car* at least comprehensible. Almost certainly there will be people at the opposite extreme who will refuse uploading no matter what the evidence. As long as it remains a matter of individual choice, this doesn't seem to be a problem.

We undergo discontinuities of consciousness every night in the ordinary world. In normal experience, the continuity of the physical body establishes identity, but we don't have any alternative to compare it with in formulating our expectations. Nanotechnology will make Moravec's scenario possible. My own expectation is that if it does work as envisioned, the intuition will change after examples become numerous and everybody knows one personally.

Uploading offers another way into a bigger world. As wide-open as the physical possibilities are with nanotechnology, they are wider still uploaded. The current-day philosopher asks, "What is it like to be a bat?"[13] but the upload could *know*. We could have new senses, not merely mapped onto our current set, and new forms of intuition, maybe even new emotions, more appropriate to the world we live in. I mentioned before how our present artificial environment has out-stripped our native equipment evolved on the African savanna; how much more will the world of tomorrow?

You must not think of such a world as overcomplex and confusing. It would be to us, but so would our world be confusing to *Homo erectus*. In fact, our descendants (and with a little luck, maybe even ourselves) will be more naturally comfortable and understand their environment more intuitively than we do ours today. That's because we've jacked up the complexity of our current world, but not the equipment we use to understand it; they will be able to do both.

Where does personal responsibility and independence go when people are programs running on the same ultra-megacomputer? Perhaps surprisingly, the range of options is the same, or perhaps even wider, than in the physical world. Let's consider a few cases, as widely scattered signposts to the vast terrain of possibilities.

There could be the equivalent of a processor per person, with communications channels between them, and one or more complex environment simulations for them to interact in. This would correspond to people with separate brains in the real world. This level of integration would interact well with real humans and people running on physically separate robot processors. The assumption here is that your thoughts are entirely yours, and that you could own a part of the physical world or simulated environment over which you would exert more or less exclusive control.

In a software world, it will be possible to create the equivalent of germs, fleas, lice, and ticks—the descendants of computer viruses— simply by thinking about them. The temptation will be great for the community to want to control your thoughts in fear of such things, even though people who did that would be as rare as people who deliberately spread disease today.

This is a significant concern because lowering the firewalls between people will have so many advantages in other ways. Exactly the same kinds of thing happened to people when they began living in cities: disease was a scourge, and epidemics like the Black Death could wipe out a third of the population. Yet people crowded into cities because it greatly facilitated communication and trade, the building of common infrastructure, and other economies of scale. And yes, there were plagues; but the advantages (usually) outweighed them. Indeed, living in such cities clearly made people stronger and more effective in the long run.[14]

In the physical world, technology has helped finesse the issue, with transportation, sanitation, medicine, and so forth. Nanotechnology can carry that further, for example, with skinsuits acting as biological firewalls but allowing direct personal contact. In the software world, the choices are harder. Uploading will allow things like direct transfer of thoughts and emotions, joint experience, and many modes of interaction as yet unthought-of. It will also allow not only direct monitoring of people's thoughts, but legislated changes in the structure of their minds. Given the track record of bureaucracies in the real world, the clear and present danger is that communities of uploads would quickly evolve into soulless monstrosities.

It should be noted that the ability to modify people is not limited to the uploads—it will happen all too soon in the physical world with brain-altering nanomedicine, genetic engineering, and the like. Indeed, the level of technology, both physical manipulation capabilities and the scientific understanding of what to manipulate, that will enable uploading in the first place is almost certainly capable of restructuring biological brains instead.

Luckily, soulless monstrosities won't win in the long run. They just can't seem to play nice with other soulless monstrosities. Evolution could have taken us that way, like ants, but didn't. There's too much value in the adaptable flexibility of the semiautonomous intelligence that we are. One of the challenges awaiting us as we move forward is to understand this well enough to avoid some unfortunate experiments.

With a properly defined Bill of Mental Rights, however, an upload community could be a truly marvelous place. It would be like the concentration of talent of a Hollywood or a Silicon Valley—centers of great creativity and an enormous value to humanity as a whole.

Not only doesn't everyone have to make the same choice, but each of us will have all the options, and can be biological humans, physically autonomous robots, and members of upload communities, serially or even at the same time. It will be trivial to copy yourself into different forms or many individuals of the same form, and not so trivial but possible to merge them back again. You can know the tight brotherhood of an upload community where you can literally feel your neighbor's emotions, the adventure of a millimetric ranger keeping the insects of Earth at bay, the thrill of a thousand-ton robot exploring the moons of the outer planets, and relax as a standard-issue human playing golf on a terraformed Mars.

TO THINE OWN SELF BE TRUE

> In what distant deeps or skies
> Burnt the fire of thine eyes?
> On what wings dare he aspire?
> What the hand dare seize the fire?
>
> —William Blake, "The Tiger"

Are you the same person you were ten, twenty, forty years ago? People grow and change. Literature in which the characters don't develop is considered flat and second rate. We are autogenous creatures, creating ourselves as we go along. Nanotechnology will give us the ability to improve ourselves, a noble pursuit, but one fraught with enigma and danger.

Living in a biological brain puts bounds on the variation we can achieve. In simple terms, we grow and learn, but we remain, willy-nilly, human. Even so, technologies that we adopt to extend our capabilities have changed us in ways we didn't anticipate. Simply compare the culture of today with that of a century ago. Our outlook is considerably more cosmopolitan, mostly due to transportation and communications. Our sexual mores have undergone a major transformation, due in some part to automobiles and contraceptives.

Imagine now that instead of such a minor shift in the gadgets we use, you were able to change your basic motivational structure, value system, and so forth. What if we did in fact change our perceptions so we found an office building as beautiful as a sunset? I can imagine wanting to change myself plenty of ways. If carried to an extreme, such changes, although small and clearly better at each step, could lead in the long run to people whose motives, reactions, and ways of thought we today could not recognize as human. Should we, as humans, countenance such a course, when we cannot know where it will lead?

We can no more know where nanotechnology will take us than our ancestors, chipping stones on the plains of Africa, knew it would lead to what we are now. Would you trade places with a *Homo erectus*? I would not. We've gained so many things: language, art, science, some would even say consciousness itself. What have we lost? Life then was nasty, brutish, and short compared to what we live today. Our present lives will seem as poor compared to what the future has to offer, if only we embrace the possibilities and work—with open eyes but also open minds—to make them reality.

It's all a matter of perspective. Either we have overrun our natural niche, jammed the globe, and have nothing left to hope for but to fight over the dwindling resources. Or we stand at the threshold of the universe, at the dawning of the age of true intelligence, and the human adventure is just beginning. The choice is yours.

NOTES

CHAPTER 1. WHAT IS NANOTECHNOLOGY?

1. Completed in Holland about 1633, *De Homine* (A Treatise of Man) was suppressed and finally published in 1662. Translation and commentary by Thomas Steele Hall (Cambridge, MA: Harvard University Press, 1972).

2. Erwin Schrödinger, *What Is Life? The Physical Aspect of the Living Cell* (Cambridge: Cambridge University Press, 1944), available online at http://home.att.net/~p.caimi/schrodinger.html.

3. Arthur C. Clarke, *Profiles of the Future: An Inquiry into the Limits of the Possible* (New York: Harper & Row, 1972), p. 21.

4. "Concerns That Nanotech Label Is Overused," *New York Times*, April 12, 2004, p. C2, col. 4.

5. K. Eric Drexler, "Molecular Engineering: An Approach to the Development of General Capabilities for Molecular Manipulation," *Proceedings of the National Academy of Sciences USA* 78, no. 9 (September 1981): 5275–78. At that time, proteins had not been designed from scratch, and doing so was considered an extremely difficult, if not impossible, problem. However, within a decade, it had been done. See L. Regan and W. F. DeGrado, "Characterization of a Helical Protein Designed from First Principles," *Science* 241 (1988): 976–78.

6. K. Eric Drexler, *Engines of Creation: The Coming Era of Nanotechnology* (New York: Anchor/Doubleday, 1984).

7. Actually, Norio Taniguchi, a Japanese engineer, was the first to get it into print, in conference proceedings in Tokyo in 1974. His definition stands up pretty well: ". . . 'Nano-technology' mainly consists of the processing of separation, consolidation and deformation of materials by one atom or one molecule."

8. Drexler, *Engines of Creation*, pp. 171–90.

9. K. Eric Drexler, *Nanosystems: Molecular Machinery, Manufacturing, and Computation* (New York: Wiley, 1992).

10. A good, readable, condensed overview of the state of current knowledge on the subject is Richard Leakey, *The Origin of Humankind* (New York: Basic Books, 1994), particularly chapter 5.

CHAPTER 2. A HANDLE ON THE FUTURE

1. Gerard O'Neill, *2081: A Hopeful View of the Human Future* (New York: Simon & Schuster, 1981), p. 157.

2. Arthur C. Clarke, *Profiles of the Future: An Inquiry into the Limits of the Possible* (New York: Harper & Row, 1972), chapters 1 and 2.

3. D. S. L. Cardwell, *Turning Points in Western Technology* (New York: SHP/Neale Watson, 1972), p. 198.

4. Aristotle's *Politics* and *Physics* can both be found at http://classics.mit.edu/Aristotle/.

5. See, e.g., Hans Moravec, *Mind Children: The Future of Robot and Human Intelligence* (Cambridge, MA: Harvard University Press, 1988), pp. 64–68; Ray Kurzweil, *The Age of Spiritual Machines* (New York: Viking, 1999), pp. 20–25.

6. K. Eric Drexler, "Rod Logic and Thermal Noise in the Mechanical Nanocomputer," in *Molecular Electronic Devices: Proceedings of the 3rd International Symposium on Molecular Electronic Devices, Arlington, Virginia, October 6–8, 1986*, ed. F. L. Carter, R. E. Siatkowski, and H. Wohltjen (North-Holland, Amsterdam: Elsevier Science, 1988), pp. 39–56.

7. See http://www.sciencemuseum.org.uk/on-line/babbage/index.asp.

8. K. Eric Drexler, "Molecular Engineering: An Approach to the Development of General Capabilities for Molecular Manipulation," *Proceedings of the National Academy of Sciences USA* 78, no. 9 (September 1981): 5275–78.

9. K. Eric Drexler, *Nanosystems: Molecular Machinery, Manufacturing, and Computation* (New York: Wiley, 1992), pp. 445–88.

10. Drexler, "Molecular Engineering," p. 78.

11. Ibid.

12. Richard Feynman, "There's Plenty of Room at the Bottom," http://www.its.caltech.edu/~feynman/plenty.html, about 65% of the way through the talk.

13. Robert A. Heinlein, *Waldo: Genius in Orbit* (New York: Avon, 1950).

14. Cardwell, *Turning Points*, pp. 122–27.

15. Bernard Jaffe, *Crucibles: The Story of Chemistry from Ancient Alchemy to Nuclear Fusion* (New York: Simon & Schuster, 1930), pp. 37–54.

16. Cardwell, *Turning Points*, p. 140.

17. Ibid. Also Terrence Kealey, *The Economic Laws of Scientific Research* (New York: St. Martin's, 1996). The "ivory tower" hadn't always been there in France, of course, or it wouldn't have attained the technological leadership it enjoyed before the industrial revolution. France's national policies and institutions underwent a sea change with the French Revolution and Napoleon.

18. National Science Foundation, http://www.nsf.gov/sbe/srs/seind00/c4/fig04-23.htm.

19. See, e.g., http://www.frankenfoods.org/.

20. This isn't strictly true. X-rays can be focused by interaction with the ever-so-slightly convergent planes of atoms in a bent piece of quartz, for example. This is sufficient to make a spectrum, like a prism does for ordinary light, but it can't be used for imaging.

21. Binnig is German; Rohrer Swiss. See http://www.nobel.se/physics/laureates/1986/.

22. Robert A. Heinlein, *Podkayne of Mars* (New York: Putnam, 1963).

CHAPTER 3. CURRENT NANOTECHNOLOGY

1. See http://www.radshield.com/.

2. See http://www.pilkington.com/international+products/activ/default.htm.

3. See http://www.zyvex.com/Products/additives.html.

4. See http://www.zyvex.com/Products/Grippers.html.

5. The original lithography, invented by Alois Senefelder ca. 1800, involved writing with a waxy ink on polished limestone that was then acid-etched around the writing to create a printing plate. *Litho* means "rock," and *graphy* means "writing."

6. When you put an electron in a small box, quantum mechanics says that you restrict the energy it can have to one of a small number of distinct amounts. This in turn restricts the light it can emit or absorb to one of a small number of wavelengths. Many atoms and molecules effectively do this to their electrons, which is why substances have the colors they have. The ability to build the boxes to whatever size you want is helpful in designing light-handling devices like lasers.

7. It's actually a bit more complicated than that. There are two strands of the DNA—the double helix—and one has C wherever the other has G, and A where it has T, and vice versa. So two "complementary" sequences of the letters are involved.

CHAPTER 4. DESIGNING AND ANALYZING NANOMACHINES

1. Ralph C. Merkle and Robert A Freitas Jr., "Theoretical Analysis of a Carbon-Carbon Dimer Placement Tool for Diamond Mechanosynthesis," *Journal of Nanoscience and Nanotechnology* 3 (August 2003): 319–24.

2. Andrew R. Leach, *Molecular Modelling: Principles and Applications* (Englewood Cliffs, NJ: Prentice Hall, 2001), pp. 165–252.

3. K. Eric Drexler, *Nanosystems: Molecular Machinery, Manufacturing, and Computation* (New York: Wiley, 1992), p. 87.

4. Leach, *Molecular Modelling*, pp. 353–400.

5. By, e.g., Nanorex, Inc.

CHAPTER 5. NUTS AND BOLTS

1. K. Eric Drexler, *Nanosystems: Molecular Machinery, Manufacturing, and Computation* (New York: Wiley, 1992), p. 25.

2. Ibid., pp. 24–28.

3. See Robert Hazen, *The Diamond Makers* (Cambridge: Cambridge University Press, 1999), for a history.

4. Quoted in Ivan Amato, *Stuff: The Materials the World Is Made Of* (New York: Basic Books, 1997), p. 154.

5. Robert A. Freitas Jr. and Ralph C. Merkle, *Diamond Synthesis and Diamond Mechanosynthesis* (Georgetown, TX: Landes Bioscience, 2005), and personal communications.

6. Carlo Montemagno (http://www.cnsi.ucla.edu/faculty/monte-magno_c.html) won the Feynman Prize (http://www.foresight.org/FI/2004 Feynman.html) for doing this.

7. Drexler, *Nanosystems*, pp. 293–300.

8. Note that this affects current technology, and life, as well: a cosmic ray can erase any given bit in your computer's memory. But the memory is built with redundancy and error correction, and messed-up bits get fixed up automatically. The same is true of the DNA in your cells.

9. Drexler, *Nanosystems*, pp. 336–41.

10. Ibid. pp. 191–249.

11. Ibid. p. 330.

12. Ibid. p. 329.

13. Ibid. p. 374.

14. Brian Wowk, "Phased Array Optics," in *Nanotechnology: Research and Perspectives*, ed. B. C. Crandall (Cambridge, MA: MIT Press, 1996), p. 147ff.

CHAPTER 6. ENGINES

1. The first, that is, that historians agree on an accurate description of, and that was widely enough known in its day to be copied and developed.

2. The chronology of steam engine development here follows D. S. L. Cardwell, *Turning Points in Western Technology* (New York: SHP/Neale Watson, 1972).

3. Ibid., pp. 84ff. Cardwell refers to 1769 as Annus Mirabilis: Watt's patent was only one of many seminal developments that year.

4. Woolf was one of the leading lights of the hotbed of engineering in Cornwall that was similar to today's Silicon Valley.

5. Sadi Carnot, *Reflections on the Motive Power of Fire* (Paris, 1824: reprint, New York: Dover, 1960). See also Henry Bent, *The Second Law: An Introduction to Classical and Statistical Thermodynamics* (New York: Oxford University Press, 1965), pp. 59–63.

6. Cardwell, *Turning Points*, p. 133.

7. Efficiency in heat engines is limited by the ratio of the temperature drop to the higher temperature, on the absolute (Kelvin) scale. But if the device isn't a heat engine, the rule simply doesn't apply.

8. K. Eric Drexler, *Nanosystems: Molecular Machinery, Manufacturing, and Computation* (New York: Wiley, 1992), pp. 397–98.

9. Ibid., pp. 421–28.

CHAPTER 7. A DIGITAL TECHNOLOGY

1. James Clerk Maxwell, *Theory of Heat* (New York: D. Appleton, 1872), pp. 308–309.

2. Note that two copies of a set of information don't represent any more information than one; making or erasing the second copy, or erasing the first while keeping the second, are reversible operations.

3. Note 12.5 million digital cameras versus 12.1 million film cameras: Photo Marketing Association.

4. No analog process could achieve the fidelity of information storage, retrieval, and replication that DNA, a digital medium, does.

5. The mechanisms by which this is done are fascinating but outside the scope of this book. The interested reader is referred to David S. Goodsell, *Our Molecular Nature: The Body's Motors, Machines, and Messages* (New York: Copernicus/Springer-Verlag, 1996); if he whets your apetite, dive into Bruce Alberts et al., *Molecular Biology of the Cell*, 4th ed. (New York: Garland, 2002), all 1,463 pages of it.

CHAPTER 8. SELF-REPLICATION

1. K. Eric Drexler, *Engines of Creation: The Coming Era of Nanotechnology* (New York: Anchor/Doubleday, 1984), pp. 55–58. See also Robert A. Freitas Jr. and Ralph C. Merkle's later assembler designs at http://www.MolecularAssembler.com.

2. They do happen at random times and places, but they happen only at precise places on the objects being built.

CHAPTER 9. FOOD, CLOTHING, AND SHELTER

1. Cooking changes the molecular structure of food, as well as making it warmer. There's no need to create the uncooked structure and change it; create it the way you need it.

2. This is one reason reversibility is so important in engines. Feed power and water into a nanoengine, and it produces hydrogen and oxygen. Feed the hydrogen and oxygen back in, and you get power and water. The closer it is to being reversible, the more of your original power you get back. It seems very likely that the efficiency of this process could exceed 99%.

3. "Was this the face that launched a thousand ships, And burnt the topless towers of Ilium?" Christopher Marlowe, *Doctor Faustus*, Project Gutenberg e-text no. 779.

CHAPTER 10. ECONOMICS

1. Adam B. Jaffe and Josh Lerner, *Innovation and Its Discontents: How Our Broken Patent System Is Endangering Innovation and Progress, and What to Do about It* (Princeton, NJ: Princeton University Press, 2004).

2. Note in passing that although patents are supposed to be for the purpose of revealing the techniques in an invention, they are almost always written by lawyers in very obscure language to try to cover as many as yet unthought-of developments as possible, making them a very poor engineering resource.

3. See chapter 15, "Runaway Replicators."

4. "Expert: Microsoft Dominance Poses Security Threat," CNN, February 2004, http://www.cnn.com/2004/TECH/biztech/02/16/microsoft.monoculture.ap/; "Microsoft Monoculture Allows Virus Spread," *New Scientist* (September 2003), http://www.newscientist.com/news/news.jsp?id=ns99994203. The *Wired* magazine article "Microsoft's War on Bugs" (http://www.wired.com/wired/archive/12.09/view.html?pg=3) begins:

These days, every Windows computer is a war zone of viruses, Trojans, spyware, and other malicious code trying to exploit security holes in Internet Explorer. One of the scariest of all, Download.Ject, discovered in late June, worked to log keystrokes (usernames, passwords, PINs). All this despite Bill Gates' 2002 declaration that security is his top priority. We asked Stephen Toulouse, Microsoft's security program manager, if Redmond is fighting a war it can't win.

5. Personal experience: the author was a systems programmer on both IBM and non-IBM machines in the 1970s.

6. The robots, of course, are the "means of production," that is, capital goods. Much of the ideological strife, including wars, in the twentieth century can be chalked up to differences of opinion on this question: Capitalism vs. Socialism vs. Communism.

7. Compare http://www.digitalhistory.uh.edu/historyonline/us25.cfm and US Census Bureau, *The Statistical Abstract of the United States*, http://www.census.gov/statab/www/, Agriculture section.

8. US Census Bureau, *Statistical Abstract*, table 558.

9. Hans Moravec, *Robot: Mere Machine to Transcendent Mind* (New York: Oxford University Press, 1999), pp. 127–38.

10. This doesn't seem very likely, does it? But suppose it were as easy to copy a robot as it is today to copy a music file from one PC to another. Then the only thing that would stop it happening would be an organized campaign against it.

11. Abraham Maslow, *Motivation and Personality* (New York: Harper & Row, 1970).

12. The real interest rate is the apparent interest rate minus the rate of inflation. It measures how much wealth is created over time by investment.

13. That doesn't mean that taxes would disappear, of course: plenty of taxes now are unnecessary. If you pay a state income tax, for example, note that Alaska, Florida, Nevada, New Hampshire, South Dakota, Tennessee, Texas, Washington, and Wyoming somehow seem to get by without one.

14. See chapter 16, "Real Dangers."

CHAPTER 11. TRANSPORTATION

1. This statement by Ford is widely quoted, including on the Ford Web site at http://media.ford.com/newsroom/release_display_new.cfm?release =81. It appears to have been issued in connection with the introduction of the Model T in 1908. Ford practiced what he preached, not only in building affordable cars, but he was an avid outdoorsman. He invented the charcoal briquette for cooking while enjoying "God's great open spaces."

2. "Carjacking" doesn't count—you're more likely to be mugged on foot than in a car, so being in the car makes you safer, not less so. Individual cars are not tempting terrorist targets. Ironically, flying increases your chances of being carjacked, since it forces you to enter and leave your car in large, public, often deserted lots, instead of being able to drive directly from one safe private place to another.

3. See http://www.moller.com/skycar/.

4. A marvelously well-written book for the general reader on the subject is Henk Tennekes, *The Simple Science of Flight: From Insects to Jumbo Jets* (Cambridge, MA: MIT Press, 1996).

5. Not to mention boundary layer control, which would also be beneficial. Note that this is a slightly nonstandard use of the word *drag*.

6. US Census Bureau, *The Statistical Abstract of the United States*, http://www.census.gov/statab/www/, Transportation, table 1084.

7. Ibid., Transportation, table 1074.

8. Ibid., Transportation, table 1091 and some arithmetic.

9. It's probably worth pointing out that hydrogen would be safer as a fuel than gasoline. A nice summary of hydrogen as a fuel can be found at http://www.borderlands.com/journal/h2.htm. In contrast to common belief, the *Hindenburg* disaster was caused not so much by the hydrogen, but by the skin paint, which was an aluminum/iron oxide compound similar to thermite, used in welding. See Addison Bain, "Colorless, Nonradiant, Blameless: A Hindenburg Disaster Study," http://www89.pair.com/techinfo/Aerostation/22_1/22_1d.pdf.

10. See http://www.investopedia.com/features/industryhandbook/airline.asp.

11. K. Eric Drexler and Robert Freitas, personal communications.

CHAPTER 12. SPACE

1. Donella H. Meadows et al., *Limits to Growth* (New York: Potomac Associates, 1972). This was a study, using sophisticated mathematics together with some very questionable premises, purporting to show that Western civilization was just about to run out of natural resources.

2. Actually there's a "solar wind" of light and radiation that can be used by "solar sails." To collect any appreciable force, however, the sails need to be miles in size. In practice, a mirror would probably have a complex nanostructure, complete with maintenance robots for micrometeorite damage—and still be as thin as aluminum foil.

3. US Census Bureau, *The Statistical Abstract of the United States*, http://www.census.gov/statab/www/, Population.

4. http://www.panynj.gov/AboutthePortAuthority/PressCenter/Press-Releases/PressRelease/index.php?id=55.

5. See Gerard K. O'Neill, *The High Frontier* (New York: Morrow, 1977), p. 134; Greg Klerkx, *Lost in Space: The Fall of NASA and the Dream of a New Space Age* (New York: Pantheon, 2004), pp. 77–78, 158; Harry Stine, *Halfway to Anywhere: Achieving America's Destiny in Space* (New York: Evans, 1996), p. 26; James Van Allen, "Appendix E: Economics of the Space Shuttle," in *Space Science Research in the United States*, http://www.wws .princeton.edu/cgi-bin/byteserv.prl/~ota/disk3/1982/8226/822611.pdf.

6. See, e.g., http://www.freerepublic.com/focus/f-news/1071333/posts and http://www.philanthropyroundtable.org/magazines/1998/november/bailey.htm.

7. See Alan C. Tribble, *The Space Environment: Implications for Spacecraft Design* (Princeton, NJ: Princeton University Press, 2003), for details of what needs to be overcome, and Robert A. Freitas Jr., *Nanomedicine, Vol I: Basic Capabilities* (Austin, TX: Landes Bioscience, 1999), for the details of the basic capabilities to be expected from nanomedicine.

8. K. Eric Drexler, *Engines of Creation: The Coming Era of Nanotechnology* (New York: Anchor/Doubleday, 1984), pp. 90–92.

9. If you like to sleep in the nude, fine—the point is that you can comfortably wear the suit inside, not that you have to.

10. O'Neill, *The High Frontier*, pp. 159–60. Note that the entire archives of the L5 News are online at http://www.l5news.org/.

11. Richard Feynman, in his appendix to the *Challenger* inquiry report, noted that space shuttle reliability estimates varied as a function of their distance from NASA management. For the main engine, for example, NASA management estimated one failure per 100,000 flights; Rocketdyne, the engine's manufacturer, put it at one in 10,000 flights; an independent NASA consulting engineer calculated 1 in 50. See Klerkx, *Lost in Space*, p. 163.

12. Consider: If a rocket has exactly the same thrust as weight, it will hover motionless in the air and go nowhere. If it has twice as much thrust as weight, it will accelerate up at 1 G (i.e., gaining 22 mph each second). It will reach one mile per second in 165 seconds, spending, let us call it, one unit of fuel per second for a total of 165 units. Now consider a rocket that uses 5 units of fuel per second and gets five times the thrust (ten times its weight); it now accelerates up at 9 G, reaching a mile per second in about 18 seconds. It's only used 5 times 18 equals 90 units of fuel, as opposed to 165. If you're fighting gravity, higher acceleration is more efficient.

13. See http://www.space.com/spacewatch/space_junk.html.

CHAPTER 13. ROBOTS

1. Iassc Asimov, *I, Robot* (New York: Doubleday, 1950).
2. "2003 AAAI Robot Competition and Exhibition," *IEEE Computational Intelligence Bulletin* 3, no. 1 (February 2004): 5–6, available online at http://www.comp.hkbu.edu.hk/~cib/2004/Feb/cib_vol3no1_confreport2.pdf.
3. J. Storrs Hall, "Utility Fog: The Stuff That Dreams Are Made Of," in *Nanotechnology: Research and Perspectives*, ed. B. C. Crandall (Cambridge, MA: MIT Press, 1996), pp. 161–84.
4. Ray Kurzweil, *The Age of Spiritual Machines* (New York: Viking, 1999), p. 145.
5. Even less detectable mixing will be possible with direct nerve taps, and will include smell and taste as well.

CHAPTER 14. ARTIFICIAL INTELLIGENCE

1. Charles Singer et al., *History of Technology*, vol. 3 (New York: Oxford University Press, 1958), p. 98. The fantail is often erroneously attributed to Andrew Meikle.
2. Norbert Wiener, *Cybernetics* (New York: Wiley, 1949).
3. Gordon J. Murphy, *Basic Automatic Control Theory* (New York: van Nostrand, 1957).
4. This story, and others like it, has attained apocryphal status by now. http://www.snopes.com/language/misxlate/machine.htm states, "How these particular tales got started is unknown, but they've been circulating for thirty or forty years now, in part because their details do accurately reflect the era in which they originated."
5. Terry Winograd, *Understanding Natural Language* (San Diego: Academic, 1972).
6. James Allen, *Natural Language Understanding* (Menlo Park, CA: Benjamin-Cummings, 1987).
7. Rodney Brooks, *Flesh and Machines: How Robots Will Change Us* (New York: Pantheon, 2002), pp. 187–91.
8. John Holland, *Adaptation in Natural and Artificial Systems: An Introductory Analysis with Applications to Biology, Control, and Artificial Intelligence* (Ann Arbor: University of Michigan Press, 1975); Marvin Minsky and Seymour Papert, *Perceptrons* (Cambridge, MA: MIT Press, 1969).
9. See, e.g., Moshe Sipper, *Machine Nature: The Coming Age of Bioinspired Computing* (New York: McGraw-Hill, 2002).
10. Sholom Weiss and Casimir Kulikowski, *Computer Systems That Learn* (San Mateo, CA: Morgan Kaufmann, 1991), pp. 108–10. The "back-propa-

gation" method is a way of letting corrections "seep" through the network, slowly adjusting each neuron.

11. Hans Moravec, *Robot: Mere Machine to Transcendent Mind* (New York: Oxford University Press, 1999), pp. 32–40. Moravec's robots make a map and calculate the probability, for each point on the map, that an object is there, given the sensor readings the robot gets.

12. Melanie Mitchell, *Analogy-Making as Perception: A Computer Model* (Cambridge, MA: MIT Press, 1993).

13. Heard in person by the author at the AAAI Fall Symposium, November 2–4, 2001, North Falmouth, MA.

14. Marvin Minsky, *The Society of Mind* (New York: Simon & Schuster, 1985), p. 245.

15. David Gelernter, *The Muse in the Machine: Computerizing the Poetry of Human Thought* (New York: Free Press, 1994), pp. 131–36.

16. As you might expect, estimates of the processing power needed for human-level AI vary considerably. I find Moravec's estimate (in *Mind Children* and *Robot*) most convincing.

17. Vernor Vinge, *The Coming Technological Singularity: How to Survive in the Post-Human Era*, http://www-rohan.sdsu.edu/faculty/vinge/misc/singularity.html.

CHAPTER 15. RUNAWAY REPLICATORS

1. First defined in K. Eric Drexler, *Engines of Creation: The Coming Era of Nanotechnology* (New York: Anchor/Doubleday, 1984), p. 172. The definitive technical treatment is Robert A. Freitas Jr., *Some Limits to Global Ecophagy by Biovorous Nanoreplicators*, http://www.foresight.org/NanoRev/Ecophagy.html.

2. See http://www.sciam.com/article.cfm?articleID=000C3AAE-D82A -10F9-975883414B7F0000 or do a Web search for Global Dimming.

3. When this was originally written, the National Nanotechnology Initiative's Web site, http://www.nano.gov/, had such a claim on its FAQ page, but it appears to have changed its mind.

4. See "The Indiana Pi Bill," http://www.agecon.purdue.edu/crd/Local gov/Second Level pages/indian _pi_bill.htm.

5. Caroline Lucas, "We Must Not Be Blinded by Science," editorial in *Guardian*, June 12, 2003, http://www.guardian.co.uk/comment/story /0,3604,975427,00.html.

6. Consider "The Threat of Eco-Terrorism," the testimony to Congress of James F. Jarboe, Domestic Terrorism Section Chief of the FBI (http://www.fbi.gov/congress/congress02/jarboe021202.htm): "During the past several years, special interest extremism, as characterized by the Animal

Liberation Front (ALF) and the Earth Liberation Front (ELF), has emerged as a serious terrorist threat." As the moderator of sci.nanotech I was flabbergasted by the expressed willingness of educated, professional, and otherwise responsible people to release replicators unilaterally to fix perceived environmental problems.

7. Blake Morrison, "Weapons Slip Past Airport Security," *USA Today*, March 25, 2002, http://www.usatoday.com/news/sept11/2002/03/25/usat-security.htm.

8. See Marcus J. Ranum, *The Myth of Homeland Security* (New York: Wiley, 2004), pp. 78–82.

CHAPTER 16. REAL DANGERS

1. Ambrose Bierce, *Devil's Dictionary* (New York: Dover, 1958), definition of "rear," p. 107.

2. See, e.g., "The Printing Press and a Changing World: Luther and the Protestant Reformation," at http://communication.ucsd.edu/bjones/Books/luther.html.

CHAPTER 17. NANOMEDICINE

1. Kenneth Walker, *The Story of Medicine* (New York: Oxford University Press, 1954), pp. 225–26.

2. Ibid., pp. 218–25.

3. Ibid., pp. 252–54.

4. Ibid., pp. 254–56.

5. Robert A. Freitas Jr., *Nanomedicine, Vol I: Basic Capabilities* (Austin, TX: Landes Bioscience, 1999), p. 15.

6. Ibid., p. 9.

7. Ibid.

8. Robert A. Freitas Jr., "Exploratory Design in Medical Nanotechnology: A Mechanical Artificial Red Cell," *Artificial Cells, Blood Substitutes, and Immobilization Biotechnology* 26 (1998): 411–30. See also http://www.foresight.org/Nanomedicine/Respirocytes.html.

9. Note that Freitas's original design calls for a spherical tank with a fixed partition. Numerous variations are possible.

10. See http://www.rfreitas.com/Nano/Microbivores.htm. Freitas describes microbivores as follows (personal communication): "Nanorobotic phagocytes (artificial white cells) called microbivores . . . could patrol the

bloodstream, seeking out and digesting unwanted pathogens including bacteria, viruses, or fungi. During each cycle of nanorobot operation, the target bacterium becomes bound to the surface of the bloodborne microbivore like a fly on flypaper, via species-specific reversible binding sites. Telescoping robotic grapples emerge from silos in the device surface, establish secure anchorage to the microbe's plasma membrane, then transport the pathogen to the ingestion port at the front of the device where the pathogen cell is internalized into a 2 [cubic] micron morcellation chamber. After sufficient mechanical mincing, the morcellated remains of the cell are pistoned into a separate 2 [cubic] micron digestion chamber where a preprogrammed sequence of forty engineered enzymes are successively injected and extracted six times, progressively reducing the morcellate to monoresidue amino acids, mononucleotides, simple fatty acids, and sugars. These basic molecules are then harmlessly discharged back into the bloodstream through an exhaust port at the rear of the device, completing the 30-second digestion cycle.

"No matter that a bacterium has acquired multiple drug resistance to antibiotics or to any other traditional treatment; the microbivore will eat it anyway, achieving complete clearance of even the most severe septicemic infections in minutes to hours, as compared to weeks or even months for antibiotic-assisted natural phagocytic defenses, without increasing the risk of sepsis or septic shock. Hence microbivores, each 2–3 microns in size, appear to be up to ~1000 times faster-acting than either unaided natural or antibiotic-assisted biological phagocytic defenses, and can extend the doctor's reach to the entire range of potential bacterial threats, including locally dense infections. Related nanorobots could be programmed to recognize and digest cancer cells, or to clear circulatory obstructions in a time on the order of minutes, thus quickly rescuing the stroke patient from ischemic damage."

11. Stephen S. Hall, *Merchants of Immortality: Chasing the Dream of Human Life Extension* (Boston: Houghton Mifflin, 2003), pp. 225–42.

12. Bruce Alberts et al., *Molecular Biology of the Cell* (New York: Garland, 2002), pp. 1344–45; David S. Goodsell, *Our Molecular Nature: The Body's Motors, Machines, and Messages* (New York: Copernicus/Springer-Verlag, 1996), pp. 38–39.

13. See http://www.gen.cam.ac.uk/sens/.

14. Freitas elaborates (personal communications): "The most likely site of pathological function in the cell is the nucleus, more specifically, the chromosomes. In one simple cytosurgical procedure, a nanorobot controlled by a physician would extract existing chromosomes from a diseased cell and insert new ones in their place. . . . The replacement chromosomes are manufactured to order, outside of the patient's body in a laboratory benchtop production device that includes a molecular assembly line, using the patient's individual genome as the blueprint. The replacement chromosomes are appro-

priately demethylated, thus expressing only the appropriate exons that are active in the cell type to which the nanorobot has been targeted. If the patient chooses, inherited defective genes could be replaced with nondefective base-pair sequences, permanently curing a genetic disease and permitting cancerous cells to be reprogrammed to a healthy state."

15. Robert Freitas, personal communication, 2004.

16. See http://www.ac.wwu.edu/~stephan/webstuff/demographs/life .data.html, taken from *Historical Statistics of the United States*.

17. US Census Bureau, *The Statistical Abstract of the United States*, http://www.census.gov/statab/www/, Health table no. 112.

CHAPTER 18. IMPROVEMENTS

1. This seems unobjectionable as long as they are under your complete control and can only interact with you. It begins to become troubling if the reading is being done by some other person. The crude beginnings of technological "mind reading" exist today, involving PET scans that determine the activity levels of the various parts of the brain. (See Bruce H. Hinrichs, "The Science of Reading Minds," *Humanist* 61, no. 3 (May/June 2001), available online at http://www.findarticles.com/p/articles/mi_m1374/is_3_61/ai _75122053. It is also worth retiterating the lesson of the Krell in the classic science fiction movie *Forbidden Planet* that it would be disastrous to allow anything other than explicit, conscious control of some kinds of assistants.

2. Bill McKibben, *The End of Nature* (New York: Random House, 1989).

3. Bill McKibben, *Enough: Staying Human in an Engineered Age* (New York: Owl/Holt, 2003), p. 34

4. Richard J. Herrnstein and Charles Murray, *The Bell Curve: Intelligence and Class Structure in American Life* (New York: Free Press, 1994), pp. 235–52.

5. McKibben, *Enough*, pp. 32–33, 43, 99–100, 162, 202–204.

6. In *Enough* McKibben mentions the following individuals: Nick Bostrom, p. 202; Damian Broderick, pp. 69, 147, 156–58, 202; Rodney Brooks, pp. 67–68, 72–75, 88, 99–100, 120–21, 162, 204–205, 225; Arthur C. Clarke, pp. 102, 107–108, 161; K. Eric Drexler, pp. 79–85, 90, 104, 113, 155, 200; Robert Freitas, pp. 90–91, 98, 147; Keith Henson, p. 120; Ray Kurzweil, pp. 68–69, 76, 81, 84, 85, 88–89, 96, 100, 117, 165, 200; John McCarthy, p. 215; Tom McKendree, p. 97; Ralph Merkle, pp. 165, 190; Marvin Minsky, pp. 96, 111, 181, 200, 203, 219; Hans Moravec, pp. 67–69, 71–72, 74, 88 93, 99–100, 117, 155, 200; and Vernor Vinge, pp. 101–102.

7. McKibben, *Enough*, pp. 1–3.

8. Arthur C. Clarke, *Profiles of the Future: An Inquiry into the Limits of the Possible* (New York: Harper & Row, 1972), pp. 113–22.

9. For a more in-depth look at how these issues are already beginning to be important, see the Center for Cognitive Liberty and Ethics at http://www.cognitiveliberty.org/.

10. William Calvin, *The Ascent of Mind: Ice Age Climates and the Evolution of Intelligence* (New York: Bantam, 1991), pp. 173–96.

11. In the sense of Susan Blackmore, *The Meme Machine* (New York: Oxford University Press, 1999).

CHAPTER 19. THE HUMAN PROSPECT

1. Xiaolan Gerasoulis, personal communication.

2. Edward O. Wilson, *The Future of Life* (New York: Vintage, 2002), p. 27.

3. US Census Bureau, *Statistical Abstract of the United States: Population* (Washington, DC: Government Printing Office, 2002), p. 8.

4. Ibid., p. 37

5. How can you retire if you're going to live forever? The trick is you don't live on what you've saved, but on the interest earned by what you've saved. In a nanotech economy, as we've seen, the productivity rates of capital can be significantly higher than they are now, so people might be able to retire after just a few years' work. More likely, if everyone's IQ is raised, we'll all continue enjoying creative work when and as we wish, neither running in the rat race nor completely quitting.

6. Ibid., p. 9.

7. See R. W. Bussard and R. D. DeLauer, *Fundamentals of Nuclear Flight* (New York: McGraw-Hill, 1965). NERVA was a design that involved heating a reaction gas with a nuclear reactor onboard the vehicle.

8. See George Dyson, *Project Orion: The True Story of the Atomic Spaceship* (New York: Henry Holt, 2002). ORION was a design for a craft pushed along by small nuclear explosions set off behind it.

9. Freitas calls this the "hypsithermal limit": the maximum release of at most a million gigawatts that can be permitted by all active nanomachinery on Earth. See his *Nanomedicine, Vol I: Basic Capabilities* (Austin, TX: Landes Bioscience, 1999), p. 175.

10. Robert Higgs, ed., *Hazardous to Our Health? FDA Regulation of Health Care Products* (Oakland, CA: Independent Institute, 1995).

11. Freeman Dyson, *Disturbing the Universe* (New York: HarperCollins, 2001).

12. Hans Moravec, *Mind Children: The Future of Robot and Human Intelligence* (Cambridge, MA: Harvard University Press, 1988), pp. 109–10.

13. Thomas Nagel, "What Is It Like to Be a Bat?" *Philosophical Review* 83 (1974): 435–50.

14. Jared Diamond, *Guns, Germs, and Steel: The Fates of Human Societies* (New York: Norton, 1996).

BIBLIOGRAPHY

Alberts, Bruce, Alexander Johnson, Julian Lewis, Martin Raff, Keith Roberts, and Peter Walter. *Molecular Biology of the Cell.* 4th ed. New York: Garland, 2002. The leading introductory college-level text.

Albus, James, and Alexander Meystel. *Engineering of Mind: An Introduction to the Science of Intelligent Systems.* New York: Wiley, 2001. Albus, of the National Institute of Standards and Technology, is one of the nation's leading roboticists. This book is a compendium of robotics systems engineering, and includes an architecture based on hierarchical feedback loops.

Allen, James. *Natural Language Understanding.* Menlo Park, CA: Benjamin-Cummings, 1987. A classic in the field.

Amato, Ivan. *Stuff: The Materials the World Is Made Of.* New York: Basic Books, 1997. One of the best popular introductions to materials science.

Arbib, Michael A. *The Metaphorical Brain: An Introduction to Cybernetics as Artificial Intelligence and Brain Theory.* New York: Wiley-Interscience, 1972. An exploratory, somewhat philosophical investigation.

Ashby, W. Ross. *An Introduction to Cybernetics.* New York: Wiley, 1963.

Asimov, Isaac. *I, Robot.* New York: Doubleday, 1950. Asimov was perhaps the prime example of a practicing scientist turned science writer, but he is most remembered today for his science fiction about robots, and his "Three Laws of Robotics."

Atkins, Peter. *Atkins' Molecules.* Cambridge: Cambridge University Press, 2003. A nice, popularly written introduction to the chemical world, with lots of pictures.

Aunger, Robert. *Darwinizing Culture: The Status of Memetics as a Science.* New York: Oxford University Press, 2000. We are what we think, and we think what we hear.

Avallone, Eugene A., and Theodore Baumeister III, eds. *Marks' Standard Handbook for Mechanical Engineers.* New York: McGraw-Hill, 1916. True nanotechnology will be closer to mechanical engineering than to chemistry.

Axelrod, Robert. *The Evolution of Cooperation.* New York: Basic Books, 1984. A classic little study of how evolution can produce cooperation in spite of the Prisoner's Dilemma.

Barr, Avron, and Edward Feigenbaum, eds. *The Handbook of Artificial Intelligence.* Palo Alto, CA: Kaufmann, 1981. An overview of a broad selection of AI systems from the 1970s.

Bent, Henry A. *The Second Law: An Introduction to Classical and Statistical Thermodynamics.* New York: Oxford University Press, 1965. Basic college thermodynamics text.

Bierce, Ambrose. *The Devil's Dictionary.* 1911. Reprint, New York: Dover, 1958.

Blackmore, Susan. *The Meme Machine.* New York: Oxford University Press, 1999. Probably the best book-length explanation of the ideas of memetics.

Bloom, Howard. *The Lucifer Principle.* New York: Atlantic Monthly Press, 1995. An examination of the dynamics of rising and falling human societies. Bloom began to look prescient after September 2001.

Braitenberg, Valentin. *Vehicles: Experiments in Synthetic Psychology.* Cambridge, MA: MIT Press, 1984. This little gem is a marvelously whimsical exploration of possible engineering ideas for intelligent systems, with lessons from neurophysiology.

Brockman, John, ed. *The Next Fifty Years: Science in the First Half of the Twenty-first Century.* New York: Vintage, 2002. Basic futurism. Brockman is the founder of *Edge* (www.edge.org).

Bronowski, Jacob. *A Sense of the Future: Essays in Natural Philosophy.* Cambridge, MA: MIT Press, 1977. A much more philosophical futurism.

Brooks, Rodney. *Flesh and Machines: How Robots Will Change Us.* New York: Pantheon, 2002. Brooks, head of MIT's AI lab, is one of the leading roboticists in the country, and also (see Albus) favors an architecture based on hierarchical feedback loops.

Buchanan, R. A. *The Power of the Machine: The Impact of Technology from 1700 to the Present.* London: Penguin, 1992. Basic history of technology.

Bussard, R. W., and R. D. DeLauer. *Fundamentals of Nuclear Flight.* New York: McGraw-Hill, 1965. Possibly the only full-fledged, comprehensive textbook in the field.

Calvin, William H. *The Ascent of Mind: Ice Age Climates and the Evolution of Intelligence.* New York: Bantam, 1991. A neurophysiologist looks at how we got here.

Cardwell, D. S. L. *Turning Points in Western Technology*. New York: SHP/Neale Watson, 1972. A very readable examination of the industrial revolution.

Carnot, Sadi. *Reflections on the Motive Power of Fire*. Paris, 1824. Reprint, New York: Dover, 1960. This slim volume founded the science of thermodynamics.

Clark, Andy. *Natural-Born Cyborgs: Minds, Technologies, and the Future of Human Intelligence*. New York: Oxford University Press, 2003. Clark is a great example of a practicing scientist who writes for the general reader.

Clark, Andy, and Josefa Toribio, eds. *Cognitive Architectures in Artificial Intelligence*. New York: Garland, 1998. A collection of technical papers on the subject.

Clarke, Arthur C. *Profiles of the Future: An Inquiry into the Limits of the Possible*. London: Victor Gollancz, 1961. Reprint, New York: Harper & Row, 1972. The definitive work of technological forecasting of the mid-twentieth century. Page references in the notes are to the 1972 hardcover edition.

Cleland, Andrew N. *Foundations of Nanomechanics: From Solid-State Theory to Device Applications*. Berlin: Springer-Verlag, 2003. Basic physics and studies of specific systems.

Crandall, B. C., ed. *Nanotechnology: Molecular Speculations on Global Abundance*. Cambridge, MA: MIT Press, 1996. A collection of articles about nanotech applications, including one by the current author on Utility Fog.

Crandall, B. C., and James Lewis. *Nanotechnology: Research and Perspectives*. Cambridge, MA: MIT Press, 1992. The proceedings of the first Foresight Conference on Nanotechnology.

Derbyshire, John. *Prime Obsession: Bernhard Riemann and the Greatest Unsolved Problem in Mathematics*. New York: Plume/Penguin, 2004. (Source for the Hilbert quote.)

Diamond, Jared. *Guns, Germs, and Steel: The Fates of Human Societies*. New York: Norton, 1996. A Pulitzer Prize–winning examination of why some human societies advance while others do not.

Drexler, K. Eric. *Engines of Creation: The Coming Era of Nanotechnology*. New York: Anchor/Doubleday, 1984. The text of this book is available online at http://www.foresight.org/EOC/.

———. *Nanosystems: Molecular Machinery, Manufacturing, and Computation*. New York: Wiley, 1992. This remains the most comprehensive technical analysis of nanomechanical engineering.

Drexler, K. Eric, Chris Peterson, and Gayle Pergamit. *Unbounding the Future*. New York: Morrow, 1991. Popularly written follow-up to *Engines of Creation*.

Dym, Clive L., and Raymond E. Levitt. *Knowledge-Based Systems in Engineering.* New York: McGraw-Hill, 1991. The use of AI techniques in designing machines.

Dyson, George. *Project Orion: The True Story of the Atomic Spaceship.* New York: Henry Holt, 2002. The road not taken.

Feigenbaum, Edward, and Julian Feldman, eds. *Computers and Thought.* New York: McGraw-Hill, 1963. The absolute classic of AI, still in print.

Feynman, Richard. *There's Plenty of Room at the Bottom.* Presentation given to the American Physical Society at Caltech on December 25, 1959. http://www.its.caltech.edu/~feynman/plenty.html.

Ford, Kenneth, Clark Glymour, and Patrick Hayes. *Android Epistemology.* Cambridge, MA: AAAI/MIT Press, 1995. A collection of essays where robotics meets philosophy.

Freitas, Robert A., Jr. *Nanomedicine, Vol I: Basic Capabilities.* Austin, TX: Landes Bioscience, 1999. Although aimed at medical applications, this encyclopedic volume is a very good general resource for molecular engineering. See http://www.nanomedicine.com.

———. Personal communications, 2000–2004.

Freitas, Robert A., Jr., and Ralph C. Merkle. *Diamond Surfaces and Diamond Mechanosynthesis.* Georgetown, TX: Landes Bioscience, 2005. An in-depth technical study of the process of building diamonds with mechanically manipulated reactive molecules.

———. *Kinematic Self-Replicating Machines.* Austin, TX: Landes Bioscience, 2004. The text of this book is available online at http://www.Molecular Assembler.com/KSRM.htm. A landmark in the field, second only to von Neumann.

Frenkel, Daan, and Berend Smit. *Understanding Molecular Simulation: From Algorithms to Applications.* San Diego: Academic/Harcourt, 1996, 2002. One of several excellent texts in this now widely practiced field.

Gelernter, David. *The Muse in the Machine: Computerizing the Poetry of Human Thought.* New York: Free Press, 1994. Gelernter is another practicing scientist who is an excellent, readable expositor.

Gernsback, Hugo. *Ralph 142C41+: A Romance in the Year 2660.* Boston: Stratford, 1925. Reprint, Lincoln: Bison/Nebraska, 2000. Gernsback is generally regarded as the founder of American science fiction.

Goddard, William A, III, Donald W. Brenner, Sergey Lishevski, and Gerald Iafrate, eds. *Handbook of Nanoscience, Engineering, and Technology.* Boca Raton, FL: CRC Press, 2003. An in-depth treatment of current-day laboratory nanotechnology.

Good, I. J., ed. *The Scientist Speculates.* New York: Basic Books, 1962. A classic in its field; it more or less *was* the field until Brockman came along.

Goodsell, David S. *Our Molecular Nature: The Body's Motors, Machines, and*

Messages. New York: Copernicus/Springer-Verlag, 1996. Accessible explanations with pictures.

Hall, Stephen S. *Merchants of Immortality: Chasing the Dream of Human Life Extension.* Boston: Houghton Mifflin, 2003. A good popular overview of the state of biotech, particularly stem cell research.

Hazen, Robert. *The Diamond Makers.* Cambridge: Cambridge University Press, 1999.

Heinlein, Robert A. *For Us, the Living: A Comedy of Customs.* New York: Scribner, 2004. Written in 1939 and published posthumously. This didactic volume, though intended as science fiction, contains the best explanation of Social Credit Theory I've seen in print.

———. *Waldo: Genius in Orbit.* New York: Avon, 1950. Heinlein invented several technologies that were quite practical and implemented soon after he published them. One was the waterbed. Another was telerobotics, as described in *Waldo.*

Heppenheimer, T. A. *Colonies in Space.* Harrisburg, PA: Stackpole, 1977. Reprint, New York: Warner, 1978. The best of the follow-on space colonization books after *High Frontier.*

Herrnstein, Richard J., and Charles Murray. *The Bell Curve: Intelligence and Class Structure in American Life.* New York: Free Press, 1994. The controversial treatment of race is actually a minor part of this book. Ignoring that, it is a good overview of a much-neglected topic.

Hofstadter, Douglas. *Fluid Concepts and Creative Analogies: Computer Models of the Fundamental Mechanisms of Thought.* New York: Basic Books, 1995. An excellent overview of AI in the middle region between the subsymbolic and the symbolic.

Holland, John H. *Adaptation in Natural and Artificial Systems: An Introductory Analysis with Applications to Biology, Control, and Artificial Intelligence.* Ann Arbor: University of Michigan Press, 1975. The seminal work on genetic algorithms.

Jaffe, Adam B., and Josh Lerner. *Innovation and Its Discontents: How Our Broken Patent System Is Endangering Innovation and Progress, and What to Do about It.* Princeton, NJ: Princeton University Press, 2004. One reviewer sums it up: "Patents are at the heart of the process of economic growth, and the process is suffering from a powerful form of cardiac disease."

Jaffe, Bernard. *Crucibles: The Story of Chemistry from Ancient Alchemy to Nuclear Fission.* New York: Simon & Schuster, 1930.

Kealey, Terrence. *The Economic Laws of Scientific Research.* New York: St. Martin's, 1996.

Klerkx, Greg. *Lost in Space: The Fall of NASA and the Dream of a New Space Age.* New York: Pantheon, 2004.

Krummenacker, Markus, and James Lewis, eds. *Prospects in Nanotechnology:*

Toward Molecular Manufacturing. New York: Wiley, 1995. An account of the Foresight Institute's First General Conference on Nanotechnology, this is still a very good introduction to the field.

Kuhn, Thomas S. *The Structure of Scientific Revolutions*. Chicago: University of Chicago Press, 1962, 1970, 1996.

Kurzweil, Ray. *The Age of Spiritual Machines*. New York: Viking, 1999.

Leach, Andrew R. *Molecular Modelling: Principles and Applications*. Harlow, UK: Addison Wesley Longman, 1996. Reprint, Englewood Cliffs, NJ: Prentice Hall, 2001. One of the leading treatises of the field.

Leaky, Richard. *The Origin of Humankind*. New York: Basic Books, 1994.

Lessig, Lawrence. *The Future of Ideas: The Fate of the Commons in a Connected World*. New York: Vintage, 2001.

Lide, David R., ed. *CRC Handbook of Chemistry and Physics*. 82nd ed. Boca Raton, FL: CRC Press, 2001. When you visit a physical scientist for the first time, the first thing you look at is the date of his *CRC Handbook*. Mine is 1972.

Lyshevski, Sergey Edward. *Nano- and Microelectromechancial Systems: Fundamentals of Nano- and Microengineering*. Boca Raton, FL: CRC Press, 2001.

Maxwell, James Clerk. *Theory of Heat*. New York: D. Appleton, 1872. A facsimile of the first edition is New York: AMS Press, 1972. Maxwell's writing was remarkably clear and well organized; reading him is still a pleasure today.

McKibben, Bill. *Enough: Staying Human in an Engineered Age*. New York: Owl/Holt, 2003.

Mewhinney, H. *A Manual for Neanderthals*. Houston: University of Texas Press, 1957. How to make stone knives in your own backyard.

Minsky, Marvin, ed. *Semantic Information Processing*. Cambridge, MA: MIT Press, 1968. An AI classic, this is a collection of projects from the first heyday of AI.

———. *The Society of Mind*. New York: Simon & Schuster, 1985. A tantalizing compendium of thoughts on the subject of AI and cognitive psychology.

Minsky, Marvin, and Seymour Papert. *Perceptrons*. Cambridge, MA: MIT Press, 1969. What neural networks can and can't do.

Mitchell, Melanie. *Analogy-Making as Perception: A Computer Model*. Cambridge, MA: MIT Press, 1993. Mitchell's thesis about Copycat.

Moravec, Hans. *Mind Children: The Future of Robot and Human Intelligence*. Cambridge, MA: Harvard University Press, 1988. A futurist classic, primarily about robotics, but with seminal exploration of uploading.

———. *Robot: Mere Machine to Transcendent Mind*. New York: Oxford University Press, 1999.

Mulhall, Douglas. *Our Molecular Future: How Nanotechnology, Robotics, Genetics, and Artificial Intelligence Will Transform Our World*. Amherst, NY: Prometheus Books, 2002.

Murphy, Gordon J. *Basic Automatic Control Theory*. New York: van Nostrand, 1957.

Murphy, Robin R. *Introduction to AI Robotics*. Cambridge, MA: MIT Press, 2000.

Newton, David E. *Recent Advances and Issues in Molecular Nanotechnology*. Westport, CT: Greenwood, 2002.

Nicholl, Desmond S. T. *An Introduction to Genetic Engineering*. Cambridge: Cambridge University Press, 2002.

O'Neill, Gerard K. *The High Frontier*. New York: Morrow, 1977. The original and still the best space colonization book.

———. *2081: A Hopeful View of the Human Future*. New York: Simon & Schuster, 1981.

Orwell, George. *1984*. London: Secker and Warburg, 1949. Text available online at http://www.liferesearchuniversal.com/orwell.html.

Pinker, Steven. *How the Mind Works*. New York: Norton, 1997.

Poole, Charles P., Jr., and Frank J. Owens. *Introduction to Nanotechnology*. New York: Wiley, 2003.

Puddephatt, R. J., and P. K. Monaghan. *The Periodic Table of the Elements*. New York: Oxford Chemistry Series, Clarendon, 1986.

Ranum, Marcus J. *The Myth of Homeland Security*. New York: Wiley, 2004.

Regis, Ed. *Nano*. New York: Little, Brown, 1995.

Reitman, Edward A. *Molecular Engineering of Nanosystems*. Berlin: Springer-Verlag, 2000.

Russell, Stuart, and Peter Norvig. *Artificial Intelligence: A Modern Approach*. Englewood Cliffs, NJ: Prentice Hall, 1995. The leading textbook in the field.

Schmidt, Stanley, and Robert Zubrin. *Islands in the Sky: Bold New Ideas for Colonizing Space*. New York: Wiley, 1996.

Shackleford, James F., and William Alexander, eds. *CRC Materials Science and Engineering Handbook*. Boca Raton, FL: CRC Press, 2001.

Singer, Charles, E. J. Holmyard, A. R. Hall, and Trevor I. Williams, eds. *A History of Technology*. 5 vols. New York: Oxford University Press, 1958.

Sipper, Moshe. *Machine Nature: The Coming Age of Bio-inspired Computing*. New York: McGraw-Hill, 2002.

Smith, George O. *Venus Equilateral*. New York: Garland, 1975. This fictional series contains the first serious treatment of the economic effects of a "matter duplicator," equivalent to a personal manufacturing system, to the author's knowledge.

Stine, G. Harry. *Halfway to Anywhere: Achieving America's Destiny in Space*. New York: Evans, 1996.

Tennekes, Henk. *The Simple Science of Flight: From Insects to Jumbo Jets*. Cambridge, MA: MIT Press, 1996.

Tribble, Alan C. *The Space Environment: Implications for Spacecraft Design*. Princeton, NJ: Princeton University Press, 2003.

Vajk, J. Peter. *Doomsday Has Been Canceled.* Culver City, CA: Peace Press, 1978. Yet another follow-on to *The High Frontier*, more explicitly arguing that space colonization was the solution to population and environmental worries.

Vinge, Vernor. *The Coming Technological Singularity: How to Survive in the Post-Human Era.* Presented at NASA's Vision-21 Symposium, Cleveland, Ohio, March 30–31, 1993. http://www.rohan.sdsu.edu/faculty/vinge/misc/singularity.html.

von Neumann, John. *The Computer and the Brain.* New Haven, CT: Yale University Press, 1958.

———. *Theory of Self-Reproducing Automata.* Edited by A. W. Burks and published posthumously. Urbana: University of Illinois Press, 1966.

Wagar, W. Warren. *A Short History of the Future.* Chicago: University of Chicago Press, 1992.

Walker, Kenneth, *The Story of Medicine.* New York: Oxford University Press, 1954.

Watson, James D., Nancy Hopkins, Jeffrey Roberts, Joan Steitz, and Alan Weiner. *Molecular Biology of the Gene.* 4th ed. Menlo Park, CA: Benjamin-Cummings, 1987.

Webb, Barbara, and Thomas R. Consi. *Biorobotics: Methods and Applications.* Cambridge, MA: AAAI/MIT Press, 2001.

Weiss, Sholom, and Casimir Kulikowski. *Computer Systems That Learn.* San Mateo, CA: Morgan Kaufmann, 1991.

Wells, H. G. *The World Set Free.* London: Collins, 1914. Reprint, London: Hogarth, 1988. In which Wells predicted nuclear war.

Wiener, Norbert. *Cybernetics.* New York: Wiley, 1949. This densely mathematical tome was a best-seller in the early fifties.

———. *God and Golem, Inc.: A Comment on Certain Points Where Cybernetics Impinges on Religion.* Cambridge, MA: MIT Press, 1964.

Williams, Trevor I. *A Short History of Twentieth-Century Technology.* New York: Oxford University Press, 1982.

Wilson, Edward O. *The Future of Life.* New York: Vintage, 2002.

Wilson, Robert, and Frank Keil, eds. *The MIT Encyclopedia of the Cognitive Sciences.* Cambridge, MA: MIT Press, 1999. Of the books listed here, second in sheer weight only to *Molecular Biology of the Cell*.

Winograd, Terry. *Understanding Natural Language.* San Diego: Academic, 1972.

Worzel, Richard. *Facing the Future: The Seven Forces Revolutionizing our Lives.* Toronto: Stoddart, 1994.

Wright, Robert: *Non-Zero: The Logic of Human Destiny.* New York: Pantheon, 2000.

GLOSSARY

Adamantine: Made of or resembling diamond. The term *diamondoid* is often used instead.

AFM: See **Atomic force microscope**.

AI: See **Artificial intelligence**.

Albedo: The proportion of light a substance or object reflects.

Amino acid: Any one of the twenty kinds of molecules that are used as the building blocks for protein.

Analogical quadrature: If A is to B as C is to D, given A, B, and C, find D.

Android robot: A robot that resembles a human, possibly so closely as to be indistinguishable in ordinary circumstances.

Antisepsis: Curing or preventing infection by killing or avoiding the introduction of germs.

Artificial intelligence: The study or practice of constructing computer programs to imitate human mental functions, including using and understanding language, solving problems, interpreting sensory input and controlling a body in the physical world, and so forth. Any program so constructed.

Assembler: A machine containing a eutactic workspace in which it does mechanosynthesis; in other words, a machine that can make objects with atomic precision.

Associative memory: Memory, in computers or nervous systems, that retrieves items remembered by similarity to the request. Most biological memory is associative, most computer memory isn't.

Atomic: Having to do with atoms. The term, especially in *atomic power* and *atomic bomb*, has an association with things nuclear because of a lack of distinction between atomic and nuclear phenomena, particularly in the public mind, before the 1960s. However, it should not, since atomic and nuclear phenomena are more distinct than atomic and everyday phenomena.

Atomic force microscope: A scanning probe that presses against its sample to feel it.

ATP: Adenosine TriPhosphate. A molecular "fuel" used to carry energy in the cell from the mitochondria, which burn sugar and fat to generate it, to the various molecular machines in the cell where it is used.

ATP synthase: The molecular machine in the mitochondrion that makes ATP.

Autogenous: Literally, "born of itself." In the engineering sense, a technology, system, or machine that can produce everything necessary to repair, extend, or rebuild itself. Usage often implicitly assumes a substrate of available items or materials, as in "The blacksmith's tools are autogenous," when we know they cannot grow wood, leather, or blacksmiths.

Axial force: In a wheel, bearing, or shaft, a force along the axis of rotation; one that would tend to pull the wheel off the axle, for example. Cf. **radial force**.

Bearing: One of the most important machine elements, a bearing allows a shaft or axle to turn freely while supporting weight or transmitting power. Bearings are necessary in any machine where something turns; without them, the machine would seize up or burn up from friction. The development of strong, precise bearings was one of the key advances of the original industrial revolution.

Bending, bond: See **Bond stretching, bending, and torsion**.

Bond: An arrangement of electrons that tends to hold atoms together. The denser and more interlinked the bonds in a material, the stronger it is. Diamond is strong because it has a very dense pattern of bonds. A diamond rod the thickness of a pencil could support roughly a thousand tons, whereas an actual pencil of wood can be broken in your fingers.

Bond stretching, bending, and torsion: The three main components of a molecular mechanics force field, that is, a mathematical description of the forces bonds exert on the atoms they connect. Stretching describes the force along the bond; bending describes the force at right angles to the bond, but toward or away from another bond; and torsion describes a force that would tend to "twist" a bond.

Boolean functions and logic: The connection between logic and arithmetic, where "true" is 1 and "false" is 0. "And" is like multiplication and "or" is like addition. Computers use Boolean logic because it's very efficient to implement it with voltages for the 1s and 0s and transistors for the "ands" and "ors."

Brownian assembly: Also called self-assembly. The random wandering of Brownian motion can bring parts together so as to assemble them, though obviously much more slowly than doing so by direct mechanical force.

Brownian motion: When objects of less than a micron in size are suspended in water, they wander around randomly because of the thermal motion of the water molecules.

Bulk technology: Any technology that deals with molecules en masse, in unordered groups, rather than individually. This includes virtually all of current-day technology.

CAD: Computer-aided design. The use of software that automates many of the drafting, measuring, and calculating tasks of the engineer.

CAM: Computer-aided manufacturing. The direct control of production machinery by computers rather than human operators. Often combined with CAD, so that an engineer can design a part on a workstation and have it built automatically. Also means content-addressable memory, a form of associative memory for computers.

Cam: A machine part that encodes a desired motion in its shape. Another part, called a follower, touches the cam and moves in the desired pattern as the cam is moved in a simple way (e.g., turned).

Carbon: Element number 6. Carbon is light and forms very strong bonds. The bonds it forms can occur in a wide variety of geometries, giving rise to the complexity of organic chemistry.

Carnot, Sadi: French engineer, 1796–1832. Inventor of thermodynamics.

Catalyst: A structure that assists a chemical reaction but which is unchanged thereby. A loveseat for shy molecules who would not normally sit so close together.

Cell: The basic unit of the organization of life. A cell has at least these essential parts: a membrane that forms its boundary; molecular machines that metabolize nutrients; molecular machines that build molecular machines; molecular machines that copy DNA; and a quantity of DNA that contains a description of all the machines in digital form.

Clarke, Arthur C.: British engineer, futurist, and novelist, 1917– . Clarke invented the communications satellite and noted the usefulness of geosynchronous orbit for them.

Classical computer: A computer built according to the commonsense understanding of reality, in which an object can only be in one place, and a wire can have only one voltage, at a time. All real computers, as this is written, are classical computers. The mathematical foundations of computer science, including Turing machines, computability, and tractibility of problems, assume classical computers. Cf. **quantum computing**.

Classical mechanics: Physical laws that correspond to common sense and apply to everyday situations, like Newton's laws of motion and Maxwell's laws of electromagnetics. Contrast with relativity and quantum mechanics.

CMOS: Complementary metal oxide semiconductor. The technology used for most computer chips today. A logic gate in CMOS involves a connection to power, representing 1; a connection to ground, representing 0; and transistors that can connect the output to either one. There are two kinds of transistors ("complementary"): ones that turn on with a 1 and off with a 0 input, and ones that turn on with a 0 and off with a 1.

Compliance: The opposite of stiffness in a mechanical part: its ability to bend in response to a force.

Computer: A machine that can store, copy, and manipulate information in digital form. Computers can be electrical, mechanical, hydraulic, or, like the DNA and its associated mechanisms, chemical. All that is required is a set of physical states and transformations that map into a coherent logical structure.

Conservative: In engineering, designing a machine so that it will work even if you made a small mistake. For example, making supporting beams stronger than your calculations indicate they strictly need to be. In other words, leaving a margin for error.

Cosmic radiation: Outer space is filled with energetic particles such as protons and electrons, which make up the "solar wind," photons such as x-rays and gamma rays, and heavier nuclei originating from the explosions of distant stars and traveling across the galaxy. These together form a hazard to unprotected humans living in space. See also **Van Allen belts**.

Covalent bond: A bond formed by the sharing of a pair of electrons between atoms.

CPU: Central processing unit. The part of a computer that does the computing. The other main part is the memory.

CVD: Chemical vapor deposition. A way of making synthetic diamond using carbon-bearing gases (or other crystals using other gases).

Cybernetics: From the Greek for "steersman," the mathematical science of control in animals and machines. The precursor to modern engineering control theory, it did not prove useful enough in explaining other phenomena to fulfill its original, somewhat gradiose, expectations. The word has come to refer to things robotic, as in cyborg, or computer-related, as in cyberspace (something of a misapplication).

Cyborg: From *cybernetic organism*, a creature that is part human (or animal) and part machine.

Diamondoid: Made of or resembling diamond, particularly at the molecular level.

Diamondoid nanotechnology: Nanotechnology using diamondoid materials for rigid, machinelike parts. Often used to mean Drexlerian nanotechnology.

Diffusive transport: In biology and "wet" chemistry, molecules get from one reaction to another by randomly bumping around in solution. This is to be contrasted with the mechanistic transport of nanotechnology, where molecules are carried by conveyor belts or robot arms in specific, preplanned paths. See **Eutactic**.

Digital: Phenomena involving a small number of discrete states or types instead of a broad range of continuous ones. Digital machinery can operate without error, where continuous machines are always characterized by unavoidable tolerance, variation, and drift.

Drexler, K. Eric: American engineer and futurist, 1955– . Originator and developer of the concept of nanotechnology as described herein.

Drexlerian nanotechnology: Nanotechnology, as used herein. Molecular engineering and molecular manufacturing. Used popularly to contrast precise with nonprecise techniques at the nanoscale.

EBL or EBNL machine: Electron beam (nano-) lithography. A machine that uses a tightly focused electron beam as a cutting tool. Today's best can get down to about 2.5 nanometers.

Einstein, Albert: German American physicist, 1879–1955. Although Einstein is famous in the popular mind for relativity, which is relatively irrelevant to nanotechnology, he was also of considerable importance in explaining the nature of the atomic world, including explanations of Brownian motion and the photoelectric effect.

Electron: Atoms, for our purposes, consist of heavy, charged nuclei surrounded by clouds of electrons. An electron is perhaps best thought of as more like a bubble than a solid particle. Electrons fill regions of space around and between nuclei that vary in shape and size. They are responsible for almost all of the physical phenomena of the everyday world except weight: besides their flows being electricity, they determine the size, strength, color, and texture of matter, and the chemical reactions between substances.

Energy: The physicist's definition of energy varies slightly from the common concept reflected in phrases like *energy crisis*. To the molecular engineer, the former is more germane, for example in the form of a molecular mechanics potential. Beyond that, however, the popular meaning is of more interest to the general discussion. By far the largest source of energy for doing useful work available to humanity is the Sun. The fraction of this that comes to the Earth is about one part in two billion. We could not use even a significant chunk of that tiny fraction here without seriously disrupting the ecosystem.

Engineering: Designing objects and systems to be built to achieve some function or purpose. Such designed objects can be, and in fact usually are, limited to using physical effects understood by the designer.

Entropy: A measure of the amount of information a system contains. The system can be physical, as in thermodynamics, or logical, as in Shannon's original theory of communication. Suppose we have a box with a single helium atom inside, and a partition down the middle. We start with the atom on the left side. Consider the one bit of information that tells us which side the atom is on. Trapped on the left, the box, or a communications channel that only emits 0s, would contain no information. If we remove the partition and let the atom bounce around, there is information, just as in a channel that transmits a sequence of 0s and 1s we don't know in advance. The second law of thermodynamics says the information can't be erased without doing a certain amount of work, as by using a piston to push the atom into a predetermined side of the box.

Enzyme: A molecular machine, typically of biological origin and made of protein, that performs a catalytic chemical function.

Eutactic: Literally, "well-ordered." Of a molecular machine system, one where the atoms move in prescribed paths or sit in prescribed places. A protein is largely eutactic, although the places the atoms sit are more loosely defined than in an adamantine nanomachine. The water and diffusive transport in a cell are not eutactic.

Evolution: The orderly change of something over time. As used herein, a shorthand for the process of variation and selection that drives the natural evolution of life, which is a special case of the general process. Anything, from clothing styles to language to corporate managements, can evolve according to the laws of variation and selection. A good grasp of these laws is essential for a sound understanding of most complex phenomena.

Femto: A millionth of a billionth of. See **Scale**.

Feynman, Richard: The second-most famous physicist of the twentieth century, after Einstein. (His 1959 presentation, "There's Plenty of Room at the Bottom," and the prizes he offered for a microscale motor and inscription connect him to nanotechnology.) The Foresight Institute offers a continuing series of prizes in his name for work leading to nanotechnology.

Fullerene: A "buckyball," the carbon molecule C60, or a carbon nanotube. They are named after Buckminster Fuller because the balls have a structure reminiscent of a geodesic dome. Balls and tubes are essentially sheets of graphite rolled into spheres or cylinders, respectively.

Gear: A mechanical part that is a major component of speed- and torque-transforming devices for rotating shafts in mechanical systems. Macroscopic gears have teeth that are shaped to an involute curve, but nanoscale gears can operate efficiently with teeth of atomic dimension and shape because of the compliance of atomic surfaces and bonds.

Genetic algorithm: A simulation of evolution in a computer, where the fitness to survive can be judged by some arbitrary test, and the "organisms" can be any arbitrary constructs. The process produces constructs that are good at passing the test. The constructs can be machine designs, and the tests can be simulations and evaluation for a desired function.

Gray goo: A popular term for one of the more extreme runaway replicator scenarios.

Hydrogen bond: Because of the distribution of the electrons in a water molecule, the hydrogen atoms undergo an electrostatic attraction to the oxygen atoms of other water molecules, forming weak bonds of an ionic nature. These bonds cause water to be much "stickier" than it would otherwise be at the molecular level.

Industrial revolution: The period around 1800 in England, when steam power and several inventions using it changed England from an agrarian to an industrial economy.

Information theory: A branch of knowledge that sits between physics and computation that offers, among other things, a good intuitive understanding of entropy and reversibility at the molecular scale. It is based on Claude Shannon's Theory of Communication and fifty years of follow-on work.

Ion: An atom that has more, or fewer, electrons than protons, and thus carries a net electric charge.

Ionic bond: Positive and negative ions attract each other, whereas those with like charges repel. Thus an ionic crystal, such as common table salt, is formed of extremely regular patterns of alternating positive ions (sodium, in the case of salt) and negative ones (chlorine). Patterns of charge like this can be used to position molecules very precisely, if all the molecules are built correctly.

L5: The fifth Lagrangian point in the Earth-Moon system. L4 and L5 are points in the Moon's orbit that form equilateral triangles with the Earth and Moon. They are stable orbits—put a satellite anywhere else and it will tend to wander off, perturbed by the Moon's gravity. L5 was one of the places suggested for the O'Neill colonies; later studies suggested other, more complex, orbits.

Lithography: Literally, "rock writing." The term is used for a wide variety of printing processes, ranging from the lithographs you hang on your wall to the etching processes used to make microprocessors.

Macroscale: This refers to things and phenomena of ordinary size that behave the way we're used to. Cf. **microscale, nanoscale.**

Mechanics: In its technical sense, this term is roughly interchangeable with *physics*: It means the mathematical description of motion, energy, and so forth in a given domain. Generally used with a modifier indicating which domain is being talked about, for example, "celestial mechanics" and "quantum mechanics."

Mechanosynthesis: Building objects (molecules) by doing chemical reactions with reagent molecules that are moved mechanically, in planned paths, rather than bouncing around randomly as in conventional (solution) synthesis.

Microbivore: An artificial white blood cell, invented by Robert Freitas; a nanorobot that kills invading microorganisms.

Micron: One millionth of a meter, also called micrometer. The name *micron* is used to avoid confusion with the measuring instrument also called a micrometer. A bacterium is about a micron wide; a human cell, 10 microns; a human hair, 100 microns.

Microscale: This refers to things and phenomena the size of molecules, like bacteria and the transistors in current-day computers. Cf. **macroscale**, **nanoscale**.

Minsky, Marvin: One of the founding fathers of artificial intelligence.

Mitochondria: The "fuel cells" of living cells. They oxidize sugar and fat to produce ATP, which the rest of the cell uses for energy.

Moiety: A part of something. In the context of nanotechnology, part of a molecule.

Molecular beam epitaxy: A fabrication method, in use today, that involves evaporating molecules of a substance in a high vacuum to allow them to fall in precise layers on the object being formed.

Molecular dynamics: The simulation of matter at a level of description where the motions of the individual atoms are taken into account, but the activities of the electrons are abstracted into models of chemical bonds.

Molecular engineering: The practice of, and body of knowledge for, designing molecular machines.

Molecular machine: A machine whose parts are on the scale of nanometers and are designed and fabricated in such detail that each atom and bond in each part is accounted for. Although the proteins in the cell can properly be called molecular machines, we are concerned here with ones that are faster, with stiffer parts, and that operate in a vacuum rather than in water.

Molecular manufacturing: Manufacturing done by molecular machines producing products to atomically precise specifications.

Molecular mechanics: The physics of an atoms-and-bonds level specification, analysis, or simulation of matter and motion.

Molecular mill: A form of molecular manufacturing in which the materials and products flow through special-purpose manufacturing machines just as products on an assembly line do in a conventional factory. To be contrasted with molecular manufacturing with a single, or few, general-purpose nanorobotic manipulators.

Molecular nanotechnology: Nanotechnology, as used in this book. Distinguished from the less precise use of the term as a buzzword for anything that is less than 100 nanometers.

Moore, Gordon: Computer hardware pioneer and the originator of Moore's Law, which states, in one general formulation, that the power of computers doubles every year and a half.

Nanocomputer: Any computer whose components are in the nanometer range in size. Such a computer could be small enough to be carried by a nanorobot the size of a cell, and powerful enough to control its actions.

Nanolithography: Any form of writing or printing at the nanometer scale. Current techniques are much like fountain pens or rubber stamps, reduced to molecular size.

Nanomedicine: The use of nanotechnology to advance the art of medicine. Nanotechnology, operating at the molecular scale, could detect and correct problems with a precision much finer than today's techniques.

Nanometer: One billionth of a meter. There are roughly 150 carbon atoms in a cubic nanometer of diamond.

Nanopants: Usually, a humorous reference to the overabundance of hype and puffery in the marketing of the earliest products of nanoscale research.

Nanoscale: This refers to things and phenomena of the size of atoms to molecules. Cf. **macroscale, microscale.**

Nanosystem: Any system consisting of nanomechanical and/or nanoelectronic devices.

Nanotechnology: 1. A technology which is composed of, and can create, molecular machines using mechanosynthesis. 2. A looser meaning, in vogue as a buzzword, is any technology involving objects less than 100 nanometers wide.

NNI: The National Nanotechnology Initiative, an umbrella funding agency of the US government that manages roughly a billion dollars annually for nanoscience research and development.

Nucleus: 1. The positively charged core of an atom that contains almost all the atom's mass. Nanotechnology (like chemistry) treats nuclei as unchangeable. 2. The central organelle of the cell in higher forms of life, containing the DNA.

Organelle: A component of a cell, such as the nucleus or a mitochondrion.

Paradigm: The word first used by philosopher of science Thomas Kuhn to refer to the coherent set of beliefs and assumptions that constitute the given level of scientific understanding. Fundamental advances in science are taken to require a "paradigm shift" in which some of the assumptions are discarded in favor of new, more insightful ones.

Phased-array: Used in conjuntion with radar or optics, an array of antennas whose phase of transmission can be controlled, or of reception detected, and be used to focus or decode the waves emitted or received without the need for reflectors or lenses. Phased-array radar is in wide use today; phased-array optics would enable real-time holograms and flat, lenseless cameras.

Photolithography: The technique used to make microprocessors and other electronics today. Light is focused through a mask onto a chip, where it causes chemical changes in very fine patterns.

Piezoelectric material: A substance, like quartz, that generates a voltage when it is bent, and bends when a voltage is applied to it.

Protein: A long-chain molecule that is composed of amino acid subunits. These are the Tinkertoys of life: most of the working machines inside a cell are made of one or a few protein molecules.

QED: See **Quantum electrodynamics**.

Quantum computing: The just-emerging field of using some of the stranger phenomena of quantum mechanics, notably the superposition of states, to do much more computation with a given piece of machinery than could be done with a classical computer. In a classical computer, for example, with four bits of state, they could take on any of sixteen different values, but only one at a time. Four entangled qubits (quantum bits), however, could take on all sixteen values at once and thus do sixteen times as much work in some applications.

Quantum electrodynamics: The theory that put a solid theoretical basis under our understanding of how light and matter work. It is too computationally intensive for most calculations, though, and sketchier forms of quantum mechanics are usually used for practical purposes.

Quantum mechanics: The mechanics, or laws of physics, governing matter and energy at very small scales. The nature of the quantum world aroused great philosophical debate in the 1920s but is old, established, scientific fact today.

Radial force: In a wheel, bearing, or shaft, one that tends to push sideways, like the weight of a car on its wheels. Cf. **axial force**.

Respirocyte: An artificial red blood cell, invented by Robert Freitas. Instead of absorbing oxygen into hemoglobin, it pumps it into a tank, and thus stores a much greater amount in the same volume.

Ribosome: The cell's "machine shop," where all the proteins are made. Each cell has many ribosomes. The ribosome reads a "blueprint" in RNA that was copied from the "master copy" in DNA in the cell's genome. It then creates a chain of amino acid molecules, which will fold up into a functioning molecular machine.

Rigid: Normal-sized steel machine parts are rigid; they do not change shape appreciably during operation. The proteins that form nature's molecular machines are floppy; the interior of a cell resembles pasta fagioli.

Scale: In the context of engineering, the property that governs why ants are built differently than elephants.

Scanning probe: A form of microscope that images by contact, rather than reflected light, objects at the molecular scale. Not unlike an old-style phonograph needle in principle.

Scanning tunneling microscope: The first form of scanning probe microscope. It uses the fact that tunneling current, which jumps the gap between the needle and the object being measured, is extremely sensitive to the distance jumped and thus is a good way to measure the gap.

Schrödinger, Erwin: Austrian physicist, 1887–1961. One of the fathers of quantum mechanics and author of the book *What Is Life?*

Seeman, Nadrian: Current-day researcher at NYU and a leader in the field of forming designed physical structures with DNA.

Self-assembly: In common usage, means the same as Brownian assembly: "viruses self-assemble in the cell." It could also mean other things, such as a half-built robot putting the rest of itself together; so Brownian assembly is preferred if precision is desired.

Self-replication: The action of something that makes a replica of itself. In the context of engineering systems, we distinguish between replication, the making of an exact copy, and reproduction, which involves the mixing of genetic material from multiple parents for the purpose of deliberate variation of the design.

Senescence: The physiological deterioration associated with aging.

SIGMA unit: The author's term for a computational element that ranges from a cybernetics-style feedback controller to a symbolic process reminiscent of Mitchell's Copycat.

Stepper motor: A motor that, like the kind in a clock, moves in fixed increments in response to distinct pulses. Commonly used in robotics to maintain precise control over the position of joints and manipulators.

Stiffness: The magnitude of the force with which a material resists being stretched, compressed, or twisted. Technically, stress divided by strain.

STM: See **Scanning tunneling microscope**.

Strain: The amount of stretching, compressing, or twisting produced by a force.

Stress: The force on an object that would tend to stretch, compress, or twist it.

Stretching, bond: See **Bond stretching, bending, and torsion**.

Substrate: Something on which something else is built. A computer is a substrate for software; our brains are a substrate for our minds.

Superposition: In quantum mechanics, an object doesn't have to be in just one place at a time, or have just one speed. Yes, it's weird.

Synthesizer (or desktop nanofactory): A small molecular manufacturing facility for domestic use.

Telemagination: A proposed ability (and word, therefor) to form images, as if imagining them, and transmit them to other people.

Telomere: A "fuse" on the end of the DNA in human cells that shortens as the cell divides. Telomeres cause the "Hayflick limit," which says that cells can divide only a certain number of times.

Thermal vibrations: The atoms of all matter are moving constantly, the higher the temperature, the faster. At room temperature, they typically move faster than the speed of sound. In a solid, they remain confined to one spot by their bonds, and vibrate in place. These vibrations, however, cause a certain variation in the shape of an object that depends on its stiffness. One reason we want to use diamond in nanomachines is that it has a high enough stiffness to keep the thermal "waviness" of the parts to a minimum.

Thermodynamics: The science of the relationship between heat and work. All fuel-burning engines from the industrial revolution till today are bound by its laws. Any nanodevice will still, of course, be bound by the laws of thermodynamics; but we can get around those laws by building machines that are not heat engines.

Tolerance: In machining, the allowable error in the size of a part. The smaller the tolerance, the more precise the part, the faster it can run, the more slowly it will wear, and in particular, the more precise the actions of the machine made of the parts.

Torsion, bond: See **Bond stretching, bending, and torsion**.

Transhumanism: The study of technological improvements to humans.

Turing, Alan M.: One of the fathers of the computer. He was an early proponent of artificial intelligence, and he proposed a famous test (called the Turing test) in an attempt to defuse the philosophical conundrums in the question, "Can a machine think?"

Utility Fog: Invented by the author, a substance composed of microscopic robots. It could take on the actual physical properties of most ordinary substances and simulate some of the ones it can't (such as density). Utility Fog could be used either in "naive mode," where separate objects would be made of Fog; or in "Fog mode," where a whole roomful of Fog would simulate air and contents by changing the behavior of the Foglets (the individual microscopic robots) without assigning specific ones to specific objects, much as a television screen displays pictures today.

Valence: The number of covalent bonds an atom is capable of forming. Elements in the same column of the periodic table tend to have the same valence.

Van Allen belts: On Earth, we are protected from cosmic radiation by the atmosphere and Earth's magnetic field. The magnetic field traps a substantial quantity of the radiation in a doughnut-shaped region around the Earth called the van Allen belts. This region is much more radioactive than empty space beyond it.

Van der Waals force: Atoms that are not bonded covalently nevertheless have a slight attraction for one another; they are "sticky." At the same time, thay have a definite size and resist being pushed into one another. It's common, if not precise usage, to call both of these phenomena van der Waals.

Virtual reality: Any technology that can give you the impression of being somewhere you aren't. An IMAX theater is a good start; the term is intended to imply the provision of at least sight, sound, and some feel. There are two basic methods: external, where the system reproduces the physical stimuli you'd experience, and internal, where the the system injects the signals your senses would have produced into your nerves directly.

VLSI: Very large scale integration, that is, of integrated circuits. The general name of the technology of microprocessor chips (and indeed of most electronics today).

Von Neumann, John: One of the fathers of the computer, and the first researcher to do a serious analysis of self-replicating machines.

Wing warping: The original Wright brothers airplane was steered by changing the shape of its wings by pulling wires. Soon this was superseded by more rigid wings with hinged flaps.

Zero-sum game: A game in which the total winnings and losings of all the players even out. The only way one can win is if another (or the others together) loses just as much. Poker is a classic example.

INDEX